The Future of Sustainability

The Future of Sustainability

Edited by

Marco Keiner

Swiss Federal Institute of Technology, ETH Zürich

 Springer

A C.I.P. Catalogue record for this book is available from the Library of Congress.

ISBN-10 1-4020-4734-7 (HB)
ISBN-13 978-1-4020-4734-3 (HB)
ISBN-10 1-4020-4908-0 (e-book)
ISBN-13 978-1-4020-4734-3 (e-book)

Published by Springer,
P.O. Box 17, 3300 AA Dordrecht, The Netherlands.

www.springer.com

Printed on acid-free paper

Cover Photo with Courtesy from NASA, Source: NASA Head quarters, Photo Department,
Washington DC, sts 111-321-024.jpg

Printed in the Netherlands

This book is
dedicated to

Hans-Georg Raidl
Prof. Dr. Eckhardt Jungfer
Dr. Hans Kissling
Prof. Dr. Klaus Giessner

and

Prof. Dr. Willy A. Schmid

They all know why.

Contents

Acknowledgements

The editor would like to thank

– Alexander Likhotal, President, Sabine Arrobbio and Marianne Berner from Green Cross International, Geneva, Switzerland, for permission to use substantial parts of the article "A New Glasnost for the Planet", first published in *The Optimist Magazine*, Issue 1, April 2004 (www.optimistmag.org)

– David Satterthwaite from the Human Settlements Programme of the International Institute for Environment and Development (IIED), London, U.K. for granting permission to reprint the chapter "Sustainability is not enough" by Peter Marcuse, originally published in *Environment and Urbanization*, Vol. 10, No. 2, Oct 1998, pp. 103–111

– Lynne O'Hara from Chelsea Green Publishing Company, White River Junction, VT, USA, for permission to reprint the chapter "Tools for the Transition to Sustainability" from "Limits to Growth: The 30-Year Update". Copyright © 2004 Dennis Meadows

– Alice Essenpreis from Springer Science and Business Media, Heidelberg, for granting permission to reprint the revised chapter "Reflections on Sustainability, Population Growth, and the Environment–2006" by Albert Bartlett, originally published in Population & Environment, Vol. 16, No. 1, September 1994, pp. 5–35

– Rachel Sykes and Rachel Warrington, International Society for Ecology and Culture, Dartington (UK), Markus Breuer and Karin Schmitt from the Novartis Foundation, Basle, for their fruitful collaboration

– Martina Koll-Schretzenmayr, Anita Schürch, Andreas Gähwiler, Oswald Roth, and especially Arley Kim from the Institute for Spatial and Landscape Planning at ETH Zurich for editorial help, layout and critical proofreading

and

– Prof. Dr. Willy A. Schmid, Head of the Institute for Spatial and Landscape Planning at ETH Zurich, for his encouragement and support over the years

"Ce qui sauve, c'est de faire un pas. Encore un pas.
C'est toujours le même pas que l'on recommence ..."

Antoine de Saint-Exupéry, *Terre des Hommes*

Rethinking Sustainability—Editor's Introduction

MARCO KEINER

The Ambiguity of 'Sustainability'

At the end of the last Millennium, when lofty visions were as ubiquitous as talk of what mankind has accomplished and in what direction it is heading, 'sustainable development' or 'sustainability' became the theoretical basis and an increasingly important societal norm for human development worldwide. For some, sustainability is "*the* way to live in harmony with the environment." (Glasby 2002) The success of both terms—'sustainability' and 'sustainable development'—stems from underlying reflections on existential problems of mankind: increasing concern over exploitation of natural resources and economic development at the expense of environmental quality (cf Ward and Dubos 1972).

Today, the objective of sustainable development is acclaimed by almost all international organizations, national governments, and also private enterprises. This general consensus seems mainly to rest upon the vague substance of the term 'sustainability' itself, which leaves much room for interpretation (Voss 1997). For the definition of 'sustainable development' we generally refer to the 1987 Brundtland Report of the World Commission on Environment and Development (WCED 1987):

1

M. Keiner (ed.), The Future of Sustainability, 1–15.
© 2006 *Springer. Printed in the Netherlands.*

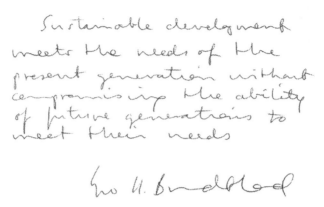

Figure 1. Definition of 'Sustainable Development'
(Autograph of Gro Harlem Brundtland)

Since the release of the Brundtland Report, this definition has been subject to several modifications and reformulated according to different points of view. Apparently, sustainable development can be easily interpreted by various groups of society according to their different interests (cf Fritsch, Schmidheiny and Seifritz 1994).

As a result, the term 'sustainable development' becomes broadly acceptable on the one hand, but on the other hand it has little specificity and loses its integrity as a political concept. The question arises whether 'sustainable development' truly represents the contemporary 'general interest'. Can one concept really form the overall framework for all policies and human activities? Isn't it only a pleonasm and politically correct selling point, since every kind of development can be more or less considered or proclaimed to be 'sustainable'? (Brunel 2004).

Today—more than ever—disagreement exists as to the precise meaning of the term. At least, 'sustainability' is 'in'. For example, the WWW search engine Google listed on July 12, 2005 the enormous number of 19.6 million hits for this term. For 'sustainable development', in turn, 17.6 million entries were found. And that only in the English language, not to mention the wealth of information to be found under 'nachhaltige Entwicklung' (German), 'desarrollo sostenible' (Spanish), 'développement durable' (French), 'desenvolvimento sustentável' (Portuguese) to name just a few translations.

Already in 1996, there were three hundred documented definitions for 'sustainability' and 'sustainable development' (Dobson 1996). Most refer to the viability of natural resources and ecosystems over time and to the maintenance of human living standards and economic growth, but even after working for two decades on coming to a common understanding of the term, its meaning remains unclear. To make matters worse, some claim that the likelihood of achieving a common understanding of 'sustainable development' is even more remote than ever (Jickling 2000).

The many definitions of 'sustainability', often general and vague, lead one to question how this norm can be of any practical value (Gremmen and Jacobs 1997). 'Sustainability' and 'sustainable development' are often misused terms; either attributed to lofty goals without a clear relationship to means or action or reduced to a catchword for business-as-usual. Today, private enterprises try to occupy the term 'sustainable development' because of its mainstream attractivity, posing an opportunity that shouldn't be missed. 'Sustainability' and 'sustainable development' are popularly used to describe a wide variety of activities which are generally ecologically laudable but which may not necessarily be sustainable in the long-term.

We should not delude ourselves into believing that we live in a sustainable world. Many ecological processes are not sustained: a broad range of species is threatened by extinction, whole ecosystems are at risk, and furthermore, climate change is becoming the most challenging threat to human life. The stability of the world as an ecosystem has been more disrupted by human activity in the last hundred years than in all of the centuries before (Gremmen and Jacobs 1997). Today, the term 'sustainable development' is not only ill defined but also misleading, because we actually live in a markedly unsustainable world, where reality is quite divorced from the vision of sustainable living, a condition that only promises to worsen in the future. There is, for example, no guarantee that our successors will survive to the year 3000 or even 2100 without some major environmental catastrophe obliterating them (Glasby 2002). By 2025 to 2030, the generation of 'baby boomers' will have retired. In the next decades, when our descendants will rule over the world, the pressure for change will unavoidably grow. And, as Jickling (2000) points out, tensions between competing interests and divergent value systems will also grow in parallel.

To survive seems to be the basic task for mankind, but going beyond sheer survivability, sustainability not only wants us to be able to survive in a hostile environment—destroyed by ourselves—but to improve living conditions for future generations (Serageldin 1996). At present, the debate on sustainable development is divided in two main opposing groups: those who argue that in order to stop self-destruction, a U-turn in human behavior and the way of our use of planet Earth needs to be implemented immediately (see, for example, Ehrlich and Ehrlich 1996); and those who believe that with new technological means human life and the condition of our planet will improve (cf Simon 1996). Perhaps, as is so often the case, the truth lies somewhere in between the two poles. At least both camps agree that some kind of transition toward more sustainability is crucial to the future of mankind.

Glasby (2002) argues that only a massive decrease in world population and resource usage phased over a century or more would permit attaining a new equilibrium that is more appropriate to a long-term occupation of planet Earth. Yet a global population even of the current size cannot adopt European and American lifestyles without destroying the environmental systems of the planet. Thus, Hughes and Johnston (2005) state, "economic growth is now increasing the world's environmental burdens much faster than population growth."

The U.S. National Research Council (1999) defines the 'sustainability transition' as a process that is possible over the next two generations, in which a stabilizing global population meets its needs, reduces hunger and poverty, and maintains Earth's life-support systems and living resources. Indeed, what is limited is the ability of deteriorating living systems to sustain a growing human population (Lovins and Lovins 2001) and also a better accessibility of the World's poor to the property mechanisms that would allow them to produce and secure greater value. It is the lack of legal property or the fact that they have no property to lose, which explains why citizens in developing and former communist states cannot conclude contracts or get credit, insurance, or utilities services (De Soto 2001).

Today, we are living through a period of rapid change and deep disturbance, having little idea in which direction we are moving, no reliable roadmap to follow, little belief in progress, and much anxiety about the dangers that lie ahead (Cowley 2003). Or, as Hales and Prescott (2002) express it:

"Making progress toward sustainability is like going to a destination we have never visited before, equipped with a sense of geography and the principles of navigation, but without a map or compass."

In conclusion, the future is more open and undetermined than our fantasy, which is conditioned by our past experiences and selective perception, can imagine (cf Dürr 1994). We are free to decide what to do with our life resources, but this freedom is always linked to a responsibility for the next generations. This is called 'intergenerational responsibility'. Moreover, as the preconditions for future development vary considerably between states as well as within states and cities, additional attention also has to be given to 'intragenerational responsibility'.

The challenge lies now in the operationalization of 'sustainable development', i.e., "the implementation of initiatives that do not merely pay lip-

service to the words but actively do justice to the original concept."
(Campbell 2000) Or, as Parris (2003) points out,

*"... defining sustainability requires a clearly articulated consensus on
what to develop, what to sustain, and for how long. It also requires
thought about how to make a transition from behaviors that trend toward
the unsustainable to ones that are more likely to be sustainable."*

If we can use this term without responsibility, if 'sustainable' just means
'lasting', does this mean that 'sustainable development' is an obsolete concept? Of course not. Jickling (2000) believes that 'sustainability' is a step-
ping-stone in the evolution of our thinking. But to this end, mankind must
evolve from *Homo sapiens* to *Homo sustinens* (cf Siebenhüner 2000) instead of the 'consume society type' of *Homo stupidus.*

The Need to Rethink 'Sustainability'

This book takes a critical look at 'sustainable development', its history and
misuse, as well as potential for future application in society. It shows logical, philosophical, and ethical reasons for reemphasizing a substantial part
of this principle and reveals several possible approaches on the levels of
political policy, economics and planning.

In a first part, the understanding and the use of 'sustainable' and 'sustainability' and their connection to 'Development' are reflected on. An
examination of major reports, carried out by **Albert A. Bartlett**, reveals
contradictory uses of the terms. In *Reflections on Sustainability, Population Growth and the Environment—2006*, Bartlett makes an attempt to give
a firm and unambiguous definition of the concept of sustainability and to
translate this definition into a series of laws, which clarify the logical
implications of the term. The laws should enable one to read the many
publications on sustainability and help decide whether the publications are
seeking to illuminate or to obfuscate. 'Sustainable development', however,
cannot take place if it is understood as 'sustainable growth', which is an
oxymoron that by definition cannot exist.

If the basic goal of development is reducing poverty, says **Herman E.
Daly** in his chapter *Sustainable Development—Definitions, Principles,
Policies*, it cannot be attained by current means (GDP growth led by global
economic integration). The obvious solution of restraining uneconomic
growth for rich countries to give opportunity for further economic growth
in poor countries is ruled out by the ideology of globalization, which can

only advocate global growth and the utility-based definition of 'sustainability'. Daly argues that we would need to promote national and international policies that charge adequately for resource rents, in order to limit the scale of the macroeconomy relative to the ecosystem and provide revenues for public purposes. These policies should be grounded in an economic theory that includes throughput among its most basic concepts. These efficient national policies would also need protection from the cost-externalizing, standards-lowering competition that is currently driving globalization.

In his chapter *Sustainability is Not Enough*, **Peter Marcuse** critically reviews the concept of sustainability, especially as it has come to be applied outside of environmental goals. It suggests 'sustainability' should not be considered as the goal of a programme—since many programmes are not sustainable—but as a constraint whose absence may limit the usefulness of a good programme. Marcuse also discusses how the promotion of 'Sustainability' may simply encourage the sustaining of the unjust status quo, i.e., the gap in wealth between post-industrialized and developing countries. He also stresses how the attempt to suggest that everyone has common interests in sustainable development masks very real conflicts of interest. One would have to consider that the costs of moving towards environmental sustainability are not borne equally by everyone, and that the definition of a 'better environment' can vary greatly. A critical analysis of how we use the term 'Sustainability' and also recognition of its limitations would be needed in order to initiate any real reform. In order to develop survivable structures, processes aiming to achieve this goal need to be flexible and adaptable to changing general conditions. Flexibility, in turn, is greatest when the number of possible options is maximized (cf Dürr 1994).

The present mainstream discourse distinguishes three basic dimensions of sustainability: economic, social and ecological, making its graphic visualization, the triple bottom line, its mantra as well as a reality for political decision-making. However, trying to direct each of the three dimensions, ecology, economy and society toward sustainable futures often results in the dilemma that the proposed solutions are incompatible with each other, e.g., that a sustainability-oriented solution for one dimension is not sustainable for another (cf Gremmen and Jacobs 1997). It has to be repeated that the concept of 'weak sustainability', in which produced assets can be substituted for natural assets, cannot lead to sustainability in a comprehensive sense. The main objectives of the three pillar model, i.e., produce more, distribute more justly, and preserve the future are hardly compatible. Thus, a solution for one dimension is only really sustainable if its effects are sustainable for the other two. However, it can be doubted that such 'ideal' solutions are achievable in a closed system like our planet as, according to the Second Law of Thermodynamics (entropy); each increase of value is at the same time accompanied by a decrease of value.

Sustainable Urban Development, Economy and Human Rights

In a second part, the book focuses on the current challenges to sustainable development, namely the phenomenon of global urbanization, economic globalization, and the role of private enterprises in regard to human rights.

The world of tomorrow will be urban. Thus, as **Marios Camhis** highlights in his chapter, *Sustainable Development and Urbanization* are closely linked issues. Until today, urbanization rates in the post-industrialized world and Latin America have passed the 75% mark and will continue to grow. Until 2030, the UN Population Division (2001) expects that up to 81% of Europeans and 85% of North Americans will live in urban areas. And, despite slower growth rates, this tendency will probably continue past 2030. On the other hand, the levels of urbanization were relatively low at the beginning of the new Millennium in less developed regions. This means, in turn, that the potential for future urban growth in developing nations is high. In Africa, the share of population living in cities will rise from 37% in 2000 to an estimated 53% in 2030, and in Asia, the same figures will mount from 48% to 54% during the same period of time. Thus in 2030, 3.8 billion people will live in urban areas in developing countries, compared to 1.4 billion in 1990 (UN Population Division 2001). This means that 80% of global growth of the urban population will take place in the poorer countries of the tropics and subtropics, and from 2000 to 2030, the urban population in developing countries will grow by 60 million people a year, effectively doubling in the period from 2000 to 2030. Until 2015, current projections foresee 27 so-called 'Mega-Cities', urban monsters with more than 10 million inhabitants.

Due to the huge scale and the multitude and complexity of problems involved, achieving sustainable development in the big cities of the developing part of the world seems to be a Sisyphean task. Focus must be laid upon the carrying capacity of planet Earth and its highly urbanized regions. In 1985, for the first time in history, the World's 'Ecological Footprint' (see Wackernagel in this book) passed the Earth's biological capacity and since then, has mounted steeply. In other terms, in 1998 the global population exceeded the Earth's carrying capacity, which is defined as the largest number of any given species (in this case, humankind) that a habitat can support indefinitely (cf Keiner 2005).

Economic globalization, a process based in the unequal and imbalanced concentration of power and distribution of resources, is responsible for the mounting environmental and social crises of the world today. **Helena Norberg-Hodge**, in her chapter *Sustainable Economies—Local or Global?*, stresses that the ostensible goal of globalization, to increase efficiency and liberalize trade, does not take into account the real costs of increased trade, which are externalized to the public through tax paid subsidies or the environment. Virtually every sphere of life is affected, from enormous investment in unsustainable infrastructures, such as transport, information and energy networks, to the loss of viability of small local businesses and diversity. Uncontrolled urbanization, environmental breakdown, economic destabilization, the erosion of democracy and government autonomy, and increased ethnic and racial conflict are only some of the true costs. Sustainable development requires a shift in direction: from globalizing economic activity towards localizing it. This does not mean that everyone must go 'back to the land', but that the forces now causing rapid urbanization should cease, reducing the unnecessary transport of goods and encouraging changes to strengthen and diversify local economies. A gradual shift towards smaller scale and more localized production would benefit both North and South and would facilitate meaningful work and more employment everywhere. Entire communities and regions would become more self-sustaining, political and economic power would be more equally distributed, and cities could regain their regional character and become more 'liveable' and less burdensome to the environment.

Public pressure has shown to be effective in bringing about changes in government policy, including resisting globalizing processes on many fronts and spawning spontaneous efforts to reweave the social and economic fabric in ways that mesh with the needs of nature. Countless numbers of such small, diverse, and local initiatives to support our local economies and communities can, if supported by policy changes over time, foster a return to long-term sustainability.

In this context, transnational corporations play a crucial role. Anderson and Cavanagh (2000) point out, that of the 100 largest economies in the world, 51 are now transnational or global corporations; only 49 are countries. The combined sales of the world's Top 200 corporations are far greater than a quarter of the world's economic activity. The Top 200 corporations' combined sales are bigger than the combined economies of all countries minus the biggest 9; that is, they surpass the combined economies of 182 countries. In other words, the Top 200 corporations have almost twice the economic clout of the poorest four-fifths of humanity.

However, these big enterprises have been net job destroyers in recent years. Their combined global employment is only 18.8 million, which is less than a third of one one-hundredth of one percent of the world's people. Not only are the world's largest corporations cutting employees, their CEOs often benefit financially from the job cuts. Big enterprises have to bear responsibility for their employees. However, very often working conditions are poor, salaries low and even the exploitation of children through child labour happens every single day. Thus, the issue of human rights in global corporations has to be addressed.

Klaus M. Leisinger, in his chapter on *Business and Human Rights*, investigates the question whether human rights are a duty for business. He firstly asks how a fair societal distribution of labour would look like. For this, he points out that modern society is differentiated into subsystems, of which he highlights the economic one. The author then discusses the societal responsibility of business firms: what they 'must', 'ought to', and 'can' do for society. Their corporate social responsibility obliges big enterprises to respect human rights, such as equal opportunity and non-discriminatory treatment, security, and appropriate working conditions. From the point of view of the corporations, he then shows the entrepreneurial options regarding these matters. In order to measure how human rights are respected in enterprises, Leisinger proposes indicators for general human rights performance to enable workers a life in dignity, justice, equality of opportunity, and fairness.

New Approaches: Global Governance, Energy Efficiency, Accounting, Evolutionability, and Transformability

Private business is one aspect; politics is another. To carry on any discussion on the future of sustainability, further questions must be asked: What has to change politically? Where are the instruments and means to implement sustainable development in everyday life and in the visioning and planning of living spaces for the coming generations?

All big institutions, like the organizations of the UN, individual States, and others are mandated and politically oriented more toward one of the objectives than to the others. For example, the International Labour Organization (ILO) or the United Nations Conference on Trade and Development (UNCTAD) do not only not collaborate with the International Monetary Fund (IMF) or The World Bank, they even denounce each other's policies. On the other side, the UNEP continues to run its environment-oriented programs following its mission to protect the environment. This sectoral splitting leads inevitably to contradictory approaches and even antagonism, resulting in confrontation and never-ending negotiations (Brunel 2004). Obviously, there is a lack of institutionalized coordination and regulation of international organizations and also a lack of comprehensive global governance (cf Stiglitz 2004). The only existing global governance today is provided by commercial and financial institutions: only the WTO, through its Dispute Settlement Body (DSB) has the mandate and the power of enforcing the rules for resolving trade quarrels, whereas for financial issues, the IMF can oblige states to modify their policies by suspending access to international financing. However, for the enforcement of norms concerning the world's 'public goods' (Samuelson 1954) health, environment, drinking water, and food, comparable global instruments or institutions do not exist. This calls for new global governance that should be devoted to the issues of long-term survival of planet Earth. The goal of achieving the sustainable use of our planet's resources will, according to Glasby (2002), take at least a century. It will require the skill, dedication and intellectual input of many people, groups of society, local and national governments, and international institutions.

Sustainable development can be seen as a simple interpretation of the general interest, an overall framework for all human activities, that guarantees everyone, anywhere and anytime, the full exercise of his rights. Thus, sustainable development cannot exist without security and liberty, and it can only be achieved if every person can satisfy its basic needs in terms of food, health, and access to education (Brunel 2004).

Despite all past disasters and doom predictions for the future for humankind and planet Earth, **Mikhail Gorbachev** claims to still be an optimist. In his chapter *A New Glasnost for Global Sustainability* he points on three principles and interlinked challenges of sustainable development: peace and security, poverty and derivation, and the environment. They are linked in terms of origin, repercussions, and the imperatives they dictate to humankind. Market-driven globalization tends to enforce the notion that economic growth determined by GNP/GDP indicators is the only way to measure national wealth and progress, and capital accumulation and individual consumption are given a higher status than social and spiritual values or cultural heritage. This kind of ideology and the policies associated with it, initiated by the countries that have benefited most from globalization, makes this trend that much stronger. The idea of 'Glasnost', or 'openness', could counteract the destructive practices associated with this kind of thinking, invigorating, informing and inspiring the citizens of the world to use our resources and knowledge for the benefit of all.

In order to change current trends, the structural factors inhibiting the transition to sustainable development need to be scrutinized. Currently prevalent behavioral patterns would need to be reversed and our value system reprioritized, adequately taking into account the relations between people and the human-nature interrelationship. A greater analysis of global issues and corresponding recommendations to politics would be needed, hence enhancing the role of science and education in our society. In addition, the media would have to act more responsibly in order to build a 'society of knowledge', collaborating with scientists to pass on important information in a credible manner.

Politics, says Gorbachev, currently lags behind the pace of change. Contemporary world politics has to grow beyond the conventional principle of balance of powers, establishing global governance based on the balance of interests that can only emerge in dialogue between cultures and civilizations and internationally recognized moral precepts.

After undergoing the dramatic and evolutionary upheaval of the Agricultural and Industrial Revolutions, society is presently heading towards the third period of profound change, the 'sustainability revolution', writes **Dennis L. Meadows** in *Tools for the Transition to Sustainability*. Our society based on material excess and consumption has reached its limits, and man's ecological footprint has once again exceeded what is sustainable, requiring the necessity for another revolution. This revolution will be organic and unplanned, arising from the visions, insights, experiments, and actions of billions of people. The key will be relevant, compelling, select information flowing in new ways to new recipients, carrying new content, and suggesting new rules and goals. This simple changing of information flows would restructure the system in a turbulent and unpredictable but inevitable way. Innovators could make the changes that transform systems, with five tools or characteristics playing an essential role: visioning, networking, truth telling, learning, and loving. Together, they will guide and motivate, joining people together to support change—innovations essential for the survival of humankind.

In *'Factor Four' and Sustainable Development in the Age of Globalization*, **Ernst Ulrich von Weizsäcker** points out that two defining milestones in global environmental awareness, the 'Limits to Growth Report' to the Club of Rome and the Earth Summit 1992 in Rio, already acknowledged the impact of high resource and energy consumption: the continuing loss of biodiversity and uncertain climate change. Current rates of development and ongoing expectations of economic growth cannot continue without technological breakthroughs. However, eco-efficiency, or slowing down the increase of labour productivity while speeding up resource productivity, could be increased by a minimum of a factor of four. In order to do so, efficiency should be made profitable through technical advances and by de-subsidizing resource use worldwide. In addition, international policy development would have to actively move towards new, more favourable framework conditions for action. Ultimately, a new technological revolution could sustain long-term profitability and sustainability only if both public and private actors played important roles in accelerating the transition.

Sustainable development is a commitment to human well-being, recognizing the reality of one diverse but ultimately finite planet. How to provide for increasing human demand while operating within the means of nature is becoming the primary challenge to make sustainable development operational. This requires both the effective management of human demand and maintenance of natural capital, including its ability to renew itself. For this task, reliable measurement tools comparing the supply of natural capital with human demand on it are indispensable. They help track progress, set targets, and drive policies for sustainability. As Hales and Prescott-Allen (2002) argue,

"For development to be sustainable, it must combine a robust economy, rich and resilient natural systems, and flourishing human communities. Rational pursuit of these goals demands that we have clear policy targets, operationalize them in terms of actions and results, devise analytical tools for deciding priority actions, and monitor and evaluate our progress."

At present, an international discussion about how indicators are able to measure the state of achievement of sustainable development is under way. If sustainability-oriented concepts are to be successful, it is essential to define measurable objectives. Thus, instruments are needed that show us whether we are making genuine progress toward or away from the context-defined targets of sustainability. The international institutions UNCSD, OECD, and the World Bank have established various frameworks for economic, social, environmental, and institutional indicators, partly differentiated into sectoral views (e.g., urban, agriculture, and so forth). An important change arising from the discussions outlined above, has been that GDP is no longer regarded as the universal measure of welfare. As GDP neither takes into account the state of the environment, natural resources and biodiversity, nor social welfare, integrated indicator sets needed to be worked out. One example is the 'Index for Sustainable Economic Welfare' (ISEW; Daly and Cobb 1989), in which consumer expenditure is balanced by such factors as income distribution and cost associated with pollution. A second example is the life cycle and product chain oriented 'Materials Intensity per Service Unit' (MIPS; Wuppertal-Institut 1993), a unit of eco-efficiency that examines the sustainability of production by quantifying the material intensity of a product or service by adding up the overall material input which humans move or extract to make that product or provide that service. Other leading assessment initiatives are, for example, the 'Human Development Report' (UNDP), 'Environmental Sustainability Index' (World Economic Forum), 'Living Planet Index' (WWF), 'Compass of Sustainability' (AtKisson & Associates), 'Dashboard of Sustainability' (Consultative Group on Sustainable Development Indicators), and the 'Barometer of Sustainability' (Prescott-Allen). In his chapter *Ecological Footprint Accounting*, **Mathis Wackernagel** describes another very popular resource measurement tool: the 'Ecological Footprint'. After explaining the assumptions involved and describing some representative findings, he provides examples of how this resource accounting tool could assist governments in managing their ecological assets and support their efforts for advancing sustainability.

Finally, one has to ask: if sustainable development is too abstract as a concept to be successfully put into practice, are there better alternatives or more appropriate models, tools or means to reemphasize the importance of the environmental, resource related aspects and establish a more workable mainstream view of sustainability? For Ignacy Sachs (1974), the concept of 'eco-development' implies establishing a hierarchy of objectives where social issues come first, secondly the environment, and only thirdly the case for economic viability without which no growth and development is possible.

In his chapter *Advancing Sustainable Development and its Implementation through Spatial Planning*, **Marco Keiner** turns to a philosophical and ethical approach on inter- and intragenerational equity and welfare to propose the 'Principle of Good Heritage', where present generations strive to create more opportunities for the future generations and leave less burdens. Reemphasizing the original purpose of 'sustainable development'—to ensure the long-term function of the world as an ecosystem and human habitat—the author then proposes the concept of 'evolutionable development' as an alternative approach. Coming back to the question on how sustainable development could be most effectively implemented, he states that the discipline of spatial planning has the mandate and the enforcement tools to do so, such as indicator based monitoring and controlling of sustainable development on the regional level. However, planning alone is not enough. Decentralization and multilevel cooperation is indispensable, as are a clear orientation of society's future visions and spatial development strategies on sustainable development, and the participation of civil society.

In *Sustainability is Dead—Long Live Sustainability*, an engaged manifesto, **Alan AtKisson** points out that human civilization is now faced with a paradox of gargantuan proportions: industrial and technological growth, the same forces that are endangering our future, should be accelerated in order to ensure it. Our technical capacities and cultural stability should be greatly enhanced while simultaneously changing almost every technological system on which we now depend so that they neither harm people nor the natural world, now or in the future. Unfortunately, denial and avoidance, and overwhelming powerlessness have been civilization's predominant responses to the warning signals coming from science and nature.

What is needed to avoid civilization's ultimate convulsion and collapse is a common sense of high purpose, bringing a critical mass of people from all walks of life and religious and cultural backgrounds together. If we cultivated a vision of ourselves as powerful and wise stewards of our planetary home, global transformation would become possible. In this sense, 'globalization' should not be viewed as the enemy as it is often portrayed, but a force to be steered, the energy harnessed, to accelerate innovations that realize a balance between the needs of people, nature's other species, and future generations of both.

References

Anderson S and Cavanagh J (2000) *Top 200: The Rise of Global Corporate Power.* Corporate Watch. www.globalpolicy.org/socecon/tncs/top200.htm, retrieved on July 15, 2005

Brunel S (2004) *Le développement durable.* Presses Universitaires de France, Paris

Campbell H (2000) Sustainable Development—Can the vision be realised? *Planning Theory & Practice,* Vol. 1, No. 2, pp. 259–284

Cowley J (2003) 12 great thinkers of our time. *New Statesman,* July 14

Daly H and Cobb J (1989) *For the Common Good.* Beacon Press, Boston

De Soto H (2001) The Mystery of Capital. *Finance & Development,* Vol. 38, No. 1. www.imf.org/external/pubs/ft/fandd/2001/03/desoto.htm, retrieved on July 14, 2005

Dobson A (1996) Environmental Sustainabilities: An analysis and a typology. *Environmental Politics,* 5(3), pp. 401–428

Dürr HP (1994) *Die 1,5-Kilowatt-Gesellschaft.* Global Challenges Network, Munich

Ehrlich PR and Ehrlich AH (1996) *Betrayal of science and reason.* Island Press, Washington DC

Fritsch B, Schmidheiny S and Seifritz W (1994) *Towards an ecologically sustainable growth society—Physical foundations, economic transitions and political constraints.* Springer, Berlin Heidelberg New York

Glasby GP (2002) Sustainable development: The need for a new paradigm. *Environment, Development and Sustainability* 4, 333–345

Gremmen B and Jacobs J (1997) Understanding sustainability. *Man and World* 30, 315–327

Hales D and Prescott-Allen R (2002) Flying blind: Assessing progress toward sustainability. In: Esty DC and Ivanova MH eds. *Global Environmental Governance: Options & Opportunities.* Yale School of Forestry & Environmental Studies, New Haven, pp. 31–52

Hughes BB and Johnston PD (2005) Sustainable futures: policies for global development. *Futures,* Vol. 37, No. 8, pp. 813–831

Jickling B (2000) A future for sustainability? *Water, Air, and Soil Pollution* 123, pp. 467–476

Keiner M (2005) Toward Gigagolis?: From urban growth to evolutionable medium-sized cities. In: Keiner M, Koll-Schretzenmayr M and Schmid WA, eds. *Managing Urban Futures—Sustainability and urban growth in developing countries.* Ashgate, Aldershot and Burlington, pp. 219–232

Lovins LH and Lovins AB (2001) Natural capitalism: Path to sustainability? *Corporate Environmental Strategy,* Elsevier, Vol. 8, No. 2, pp. 99–108

National Research Council, Board on Sustainable Development (1999) *Our Common Journey: A Transition Toward Sustainability.* National Academy Press, Washington DC

Parris TM (2003) Toward a sustainability transition: the international consensus. *Environment* 45 (Jan/Feb), pp. 12–22

Sachs I (1974) Ecodevelopment. *Ceres* 17, No. 4, pp. 17–21

Samuelson PA (1954) The Pure Theory of Public Expenditure—*The Review of Economics and Statistics,* Vol. 36, No. 4, pp. 387–389

Schmidt-Bleek F (1993) *Wieviel Umwelt braucht der Mensch—MIPS, das Mass für ökologisches Wirtschaften.* Birkhäuser, Basel, Boston, Berlin. English translation: The Fossil Makers, www.factor10-institute.org/seiten/pdf.htm, retrieved on July 15, 2005

Serageldin I (1996) Sustainability as opportunity and the problem of social capital. *The Brown Journal of World Affairs,* Vol. 3, No. 2, pp. 187–203

Siebenhüner B (2000) Homo sustinens—Towards a new conception of humans for the science of sustainability. *Ecological Economics* Vol. 32, No. 1, pp. 15–25

Simon JL (1996) *The ultimate resource 2.* Princeton University Press, Princeton

Stiglitz, JE (2004) *Governance for a sustainable world.* Edited transcript. Center for Corporate Responsibility and Sustainability at the University of Zurich, CCRS Occasional Paper Series, 01/04

UN Populations Division (2001) *The State of World Population.* www.unfpa.org, retrieved on October 3, 2003

Voss, G (1997) Das Leitbild der nachhaltigen Entwicklung—Darstellung und Kritik. *Beiträge zur Wirtschafts- und Sozialpolitik* 237 (4)

Ward B and Dubos R (1972) *Only one Earth—The Care and Maintenance of a Small Planet.* Andre Deutsch, London

WCED The World Commission on Environment and Development (1987) *Our Common Future.* Oxford University Press, Oxford

Reflections on Sustainability, Population Growth, and the Environment—2006[1]

ALBERT A. BARTLETT

In the 1980s, it became apparent to thoughtful individuals that populations, poverty, environmental degradation, and resource shortages were increasing at a rate that could not long be continued. Perhaps most prominent among the publications that identified these problems in hard quantitative terms and then provided extrapolations into the future, was the book *Limits to Growth* (Meadows et al. 1972), which simultaneously evoked admiration and consternation. The consternation came from traditional 'Growth is Good' groups all over the world. Their rush to rebuttal was immediate and urgent, prompted perhaps by the thought that the message of Limits was too terrible to be true (Cole et al. 1973). As the message of Limits faded, the concept of limits became an increasing reality with which people had to deal. Perhaps, as an attempt to offset or deflect the message of Limits, the word 'sustainable' began to appear as an adjective that modified common terms. It was drawn from the concept of 'sustained yield,' which is used to describe agriculture and forestry when these enterprises are conducted in such a way that they could be continued indefinitely, i.e., their yield could be sustained. The use of the new term 'sustainable' provided comfort and reassurance to those who may momentarily have wondered if possibly there were limits. The word was soon applied in many areas, and with less precise meaning, so that for example, with little visible change, 'development' became 'sustainable development,' etc. One would see political leaders using the term 'sustainable' to describe their goals as they worked hard to create more jobs, to increase population, and to increase rates of consumption of energy and resources. In the manner of *Alice in Wonderland*, and without regard for accuracy or consistency, 'sustainability' seems to have been redefined flexibly to suit a variety of wishes and conveniences.

M. Keiner (ed.), The Future of Sustainability, 17–37.
© 2006 *Springer. Printed in the Netherlands.*

The Meaning of Sustainability

First, we must accept the idea that 'sustainable' has to mean 'for an unspecified long period of time'.

Second, we must acknowledge the mathematical fact that steady growth (a fixed percent per year) gives very large numbers in modest periods of time. For example, a population of 10,000 people growing at 7% per year will become a population of 10,000,000 people in just 100 years (Bartlett 1978).

From these two statements we can see that the term 'sustainable growth' implies 'increasing endlessly'. This means that the growing quantity will tend to become infinite in size. The finite size of resources, ecosystems, the environment, and the Earth, lead to the most fundamental truth of sustainability:

"When applied to material things, the term 'sustainable growth' is an oxymoron."

(It is possible, on the other hand, to have sustainable growth of non-material things such as inflation.)

Daly has pointed out that 'sustainable development' may be possible if materials are recycled to the maximum degree possible and if one does not have growth in the annual material throughput of the economy (Daly 1994).

The Use of the Term 'Sustainable'

A sincere concern for the future is certainly the factor that motivates many who make frequent use of the word 'sustainable'. But there are cases where one suspects that the word is used carelessly, perhaps as though the belief exists that the frequent use of the adjective 'sustainable' is sufficient to create a sustainable society. 'Sustainability' has become big-time. University centers and professional organizations have sprung up using the word 'sustainable' as a prominent part of their names. In some cases, these big-time operations may be illustrative of what might be called the 'Willie Sutton[2] School of Research Management'.

For many years, studies had been conducted on ways of improving the efficiency with which energy is used in our society. These studies have been given new luster by referring to them now as studies in the 'sustainable use of energy.'

The term 'sustainable growth' is used by our political leaders even though the term is clearly an oxymoron. In a recent report from the Environmental Protection Agency we read that President Clinton and Vice President Gore wrote in Putting People First,

"We will renew America's commitment to leave our children a better nation—a nation whose air, water, and land are unspoiled, whose natural beauty is undimmed, and whose leadership for sustainable global growth is unsurpassed." (EPA 1993)

We even find a scientist writing about 'sustainable growth':

"... the discussions have centered around the factors that will determine [a] level of sustainable growth of agricultural production." (Abelson 1990)

And so we have a spectrum of uses of the term 'sustainable'. At one end of the spectrum, the term is used with precision by people who are introducing new concepts as a consequence of thinking profoundly about the long-term future of the human race. In the middle of the spectrum, the term is simply added as a modifier to the names and titles of very beneficial studies in efficiency, etc. that have been in progress for years. Near the other end of the spectrum, the term is used as a placebo. In some cases the term may be used mindlessly (or possibly with the intent to deceive) in order to try to shed a favorable light on continuing activities that may or may not be capable of continuing for long periods of time. At the very far end of the spectrum, we see the term used in a way that is oxymoronic.

Let us examine the use of the term 'sustainable' in some major environmental reports.

Sustainability

The terms 'sustainable' and 'sustainability' burst into the global lexicon in the 1980s as the electronic news media made people increasingly aware of the growing global problems of overpopulation, drought, famine, and environmental degradation that had been the subject of Limits to Growth in the early 1970s (Meadows et al. 1972). A great increase of awareness came with the publication of the report of the United Nations World Commission on Environment and Development, the Brundtland Report, which is available in bookstores under the title *Our Common Future* (Brundtland 1987).

In graphic and heart-wrenching detail, the Report places before the reader the enormous problems and suffering that are being experienced with growing intensity every day throughout the underdeveloped world. In the foreword, before there was any definition of 'sustainable', there was the ringing call:

"What is needed now is a new era of economic growth—growth that is forceful and at the same time socially and environmentally sustainable." (1987 p. xii)

One should be struck by the fact that here is a call for 'economic growth' that is 'sustainable'. One has to ask if it is possible to have an increase in economic activity (growth) without having increases in the rates of consumption of non-renewable resources? If so, under what conditions can this happen? Are we moving toward those conditions today? What is meant by the undefined terms, 'socially sustainable' and 'environmentally sustainable'?

As we have seen, these two concepts of 'growth' and 'sustainability' are in conflict with one another, yet the Brundtland Report calls for both. The use of the word 'forceful' would seem to imply 'rapid', but if this is the intended meaning, it would just heighten the conflict.

A few pages later in the Report we read:

"Thus sustainable development can only be pursued if population size and growth are in harmony with the changing productive potential of the ecosystem." (1987 p. 9)

One begins to feel uneasy. 'Population size and growth' are vaguely identified as possible problem areas, but we don't know what the Commission means by the phrase "in harmony with...?" It can mean anything. By page 11 the Commission acknowledges that population growth is a serious problem, but then:

"The issue is not just numbers of people, but how those numbers relate to available resources ... Urgent steps are needed to limit extreme rates of population growth."

The suggestion that "the issue is not just numbers of people" is alarming. This denial of the importance of numbers has become central to many of the programs that deal with sustainability. Neither 'limit' nor 'extreme' are defined, and so the sentence gives the impression that most population growth is acceptable and that only the undefined 'extreme rates of population growth' need to be dealt with by some undefined process of limiting. By page 15 we read that:

"A safe, environmentally sound, and economically viable energy pathway that will sustain human progress into the distant future is clearly imperative."

Here we see the recognition that energy is a major long-term problem, yet we see no recognition of the enormous technical and economic difficulties that can reasonably be expected in the search for an "environmentally sound, and economically viable energy pathway." The Report does recognize that 'sustainable' has to mean 'into the distant future'.

As the authors of the Report searched for solutions, they called for large efforts to support 'sustainable development'. The Report's definition of 'sustainable development' has been widely used by others. It appears in the first sentence of Chapter 2 (1987 p. 43):

"Sustainable development is development that meets the needs of the present without compromising the ability of future generations to meet their own needs."

This definition, coupled with the earlier statement of the need to "sustain human progress into the distant future," is crucial for an understanding of the term 'sustainable development'.

Unfortunately, the definition gives no hint regarding the courses of action that could be followed to meet the needs of the present, but which, in doing so, would not limit the ability of generations, throughout the distant future, to meet their own needs. It seems obvious that non-renewable resources consumed now will not be available for consumption by future generations.

The Commission recognizes that there is a conflict between population growth and development (1987 p. 44):

"An expansion in numbers [of people] can increase the pressure on resources and slow the rise in living standards in areas where deprivation is widespread. Though the issue is not merely one of population size, but of the distribution of resources, sustainable development can only be pursued if demographic developments are in harmony with the changing productive potential of the ecosystem."

Can the Commission mean that population growth slows the rise of living standards only "in areas where deprivation is widespread?" This statement recites again the politically correct assertion that "the issue is not merely one of population size." The Commission shifts the blame for the problems to presumed faults in the distribution of resources. The Commission then speaks of 'demographic developments', whatever that may mean, which must be "in harmony with...", whatever that means. If one accepts reports of the decline of 'global productive potential of ecosystems' due to deforestation, the loss of topsoil, pollution, etc. (Kendall and Pimentel 1994), then the "in harmony with..." could mean that population also will have to decline. But the Commission is very careful not to suggest the need for a decline in population.

These quotations are thought to be representative of the vague and contradictory messages that are in this important report. As the Report seeks to address severe global problems, it clearly tries to marginalize the role of population size as an agent of causation of these severe global problems.

The Brundtland Commission Report's discussion of 'sustainability' is both optimistic and vague. The Commission probably felt that, in order to be accepted, the discussion had to be optimistic, but given the facts, it was necessary to be vague and contradictory in order not to appear to be pessimistic.

Carrying Capacity

The term 'carrying capacity', long known to ecologists, has also recently become popular. It "refers to the limit to the number of humans the earth can support in the long term without damage to the environment." (Giampietro et al. 1992)

The concept of carrying capacity is central to discussions of population growth. The concept has been examined by Cohen (1995) in the book *How Many People can the Earth Support?* Cohen makes a scholarly examination of many past estimates of the carrying capacity of the Earth, and concludes that it is not possible to say how many people the Earth can support. Obviously, it depends on the desired average standard of living.

There is no closed formula for calculating the carrying capacity of the Earth, even for some stated average standard of living. This means that any calculated estimate of the carrying capacity of the Earth may be challenged and will certainly be ignored.

Human activities have already caused great change in the global environment. May (1993) observes,

"... the scale and scope of human activities have, for the first time, grown to rival the natural processes that built the biosphere and that maintain it as a place where life can flourish."

Many facts testify to this statement. It is estimated that somewhere between 20 and 40 percent of the earth's primary productivity, from plant photosynthesis on land and in the sea, is now appropriated for human use.

An impact on the global environment of this magnitude is properly the cause for alarm.

The inevitable and unavoidable conclusion is that if we want to stop the increasing damage to the global environment, as a minimum, we must stop population growth.

So, instead of trying to calculate how many people the Earth can support, we should instead, focus on the question of why should we have more population growth. This is nicely framed in the challenge:

> *"Can you think of any problem, on any scale,*
> *from microscopic to global,*
> *Whose long-term solution is in any demonstrable way,*
> *Aided, assisted, or advanced, by having larger populations*
> *At the local level, the state level, the national level, or globally?"*

The Final Word on the Carrying Capacity of the Earth

Even though we cannot calculate a carrying capacity for the Earth, we have an unambiguous indication that the world population has already exceeded this carrying capacity. We are observing global warming. If any part of the observed global warming is due to the activity of humans, then this is positive proof that the present population of the Earth, living as we do, is greater than the carrying capacity of the Earth.

Population and the Environmental Protection Agency

The US Environmental Protection Agency has done many constructive and beneficial things. The policies, actions, and leadership of the Agency are crucial if we are to have any hope of achieving a sustainable society. In a recent report from the Agency (EPA 1993) we read,

> *"In view of the increasing national and international interest in sustainable development, Congress has asked the Environmental Protection Agency (EPA) to report on its efforts to incorporate the concepts of sustainable development into the Agency's operations."*

The Report (1993) is both encouraging and distressing. It is encouraging to read of all of the many activities of the Agency, which help protect the environment. It is distressing to search in vain through the Report for acknowledgment that population growth is at the root of most of the problems, which the Agency seeks to address. While the Brundtland Report says that population growth is not the central problem, the EPA report avoids making this allegation. But the EPA report makes only a very few minor references to the environmental problems that arise as a direct consequence of population growth.

For example, the EPA report speaks of an initiative to pursue sustainable development in the Central Valley of California:

"... where many areas are experiencing rapid urban growth and associated environmental problems.... A stronger emphasis on sustainable agricultural practices will be a key element in any long-term solutions to problems in the area."

There is no way that a stronger emphasis on 'sustainable agricultural practices' can stop the 'rapid urban growth' that is destroying farmland! An emphasis on agriculture cannot solve the problem. To solve the problems, one must stop the 'rapid urban growth', which causes the problems. It is pointless to focus on the development of 'sustainable agricultural practices' when the Agency expects that agriculture will soon be displaced by the 'rapid urban growth'.

This quotation of a minor section of the EPA report makes it clear that the EPA understands the origin of environmental problems. Here is an agency that seeks to solve problems caused by population growth, yet when it sets forth its recommended solutions, stopping population growth is not mentioned. Is this professionally ethical?

The Marginalization of Malthus

We have seen how major national and international reports misrepresent and downplay (marginalize) the quantitative importance of the arithmetic of population sizes and growth. The recognition of the importance of quantitative analysis of population sizes was first popularized by Thomas Malthus two hundred years ago (Appleman 1976), but the attempted marginalization of Malthus goes on today at all levels of society.

In an article *The Population Explosion is Over,* Ben Wattenberg finds support for the title of his article in the fact that fertility rates are declining in parts of the world (Wattenberg 1997). Currently, most of the countries of Europe are at zero population growth or negative population growth, and fertility rates in parts of Asia, have declined dramatically. Rather than rejoicing over the clear evidence of this movement in the direction of sustainability, Wattenberg sounds the alarm over the 'birth dearth' as though this fertility decline requires an immediate reversal.

The most extreme case is that of Julian Simon who advocates continued population growth long into the future. Writing in the newsletter of a major think tank in Washington DC, Simon (1995) says,

"We have in our hands now—actually in our libraries—the technology to feed, clothe, and supply energy to an ever-growing population for the next 7 billion years... Even if no new knowledge were ever gained ... we would be able to go on increasing our population forever."

It has been noted that a spherical earth is finite, but a flat earth can be infinite in extent. So if Simon is correct, we must be living on a flat earth (Bartlett 1996).

The World's Worst Population Problem

Echoing a view expressed earlier by the Ehrlichs (Ehrlich and Ehrlich 1992) Bartlett points out that because of the high per capita consumption of resources in the US, we in the US have the world's worst population problem! (Bartlett 1997) Many Americans think of the population problem is a problem only of 'those people' in the undeveloped countries, but this serves only to draw attention away from the difficulties of dealing with our own problems here in the US. It is easier to tell a neighbor to mow his/her yard than it is for us to mow our own yard. With regard to other countries, we can offer family planning assistance on request, but in those countries we have no jurisdiction or direct responsibility. Within our own country we have complete jurisdiction and responsibility, yet we fail to act to help solve our own problem. In a speech at the University of Colorado, then US Senator Tim Wirth observed that the best thing we in the US can do to help other countries stop their population growth is to set an example and stop our own population growth here in the US.

There can be no question about the difficulty that we will have to achieve zero growth of the population of the US. An examination of the simple numbers makes the difficulty clear. In particular, population growth has 'momentum' which means that if one makes a sudden change in the fertility rate in a society, the full effect of the change will not be realized until every person has died who was living when the change was made. Thus it takes approximately 70 years to see the full effect of a change in the fertility rate (Bartlett and Lytwak 1995).

Population Growth Never Pays for Itself

There are many encouraging signs from communities around the US that indicate a growing awareness of the local problems of continued unrestrained growth of populations, because population growth in our communities never pays for itself. Taxes and utility costs must increase in order to pay for the growth. In addition, growth brings increased levels of congestion, pollution and frustration.

The positive proof that population growth does not pay for itself is seen in the budget crises of many US States. During the 1990s the economy was 'healthy', which means it was growing rapidly. If the growth had paid for itself, the state governments should have accumulated financial reserves to help get through a decline in the national economy. When the economy declined around the turn of the century, the fiscal obligations that had accumulated during the good times came due, and there were inadequate funds to meet the needs.

Fodor (1999) gives many detailed examples from communities all over the US showing how the population growth falls far short of paying for itself.

The Tragedy of the Commons (Hardin 1968) makes it clear that there will always be large opposition to programs of making population growth pay for itself. Those who profit from growth will use their considerable resources to convince the community that the community should pay the costs of growth. In our communities, making growth pay for itself could be a major tool to use in stopping the population growth.

Pseudo Solutions: Growth Management—Smart Growth

The claim is often made that 'smart growth will save the environment'. It is worth remembering that:

Smart growth is better than dumb growth, but
Smart growth destroys the environment; and
Dumb growth destroys the environment.
The difference is that smart growth
destroys the environment with good taste
So it's a little like buying a ticket on the TITANIC.
If you're smart you go first class.
If you're dumb you go steerage.
But either way, the result is about the same.

Pseudo Solutions: Regional Planning

As populations of cities grow, the call is made for 'regional solutions' to the many problems created by growth. This has two negative effects:

– Regional planning dilutes democracy. A citizen participating in public affairs has five times the impact in his/her city of 20,000 as he/she would have in a region of 100,000 people;

– The regional 'solutions' are usually designed to accommodate the predicted growth and hence these 'solutions' encourage more growth. In the spirit of Eric Sevareid's Law (below), regional 'solutions' enlarge the problems rather than solving them.

One concludes that regional solutions to problems caused by growth will make lives better for people only if the growth is stopped. If the regional solutions permit or encourage more growth, then the regional planning has made things worse.

War and Peace

At the local or state levels, there is an interesting parallel between the promotion of growth (unsustainability) and the promotion of war, both of which can be very profitable for high level people but are very expensive for everyone else.

The waging of war is the sole enterprise of large military establishments. Even the meanest mind knows what has to be done to win a war: one has to beat the opponent, after which one can have a large party to celebrate the victory, pass out the medals, and then start preparing for the next war. Promoting community growth is quite similar. The promotion of growth is the sole enterprise of large municipal and state establishments, both public and private. It does not take much of a mind to know that victory in the growth war requires that your community beat competing communities to become the location of new factories and businesses. Campaigns and battles are planned and, when a factory comes, there is a large party to celebrate the victory and pass out the awards. Then the community warriors start fighting for even more new factories.

In contrast, winning the peace is quite different. Even the best minds don't know for sure the best way to 'win the peace'. Compared to the groups that promote war, the public agencies that are devoted to maintaining peace are miniscule. In the effort to maintain peace, there is no terminal point at which a party is in order where all can celebrate the fact that

'We won the peace!' Winning the peace takes eternal vigilance. Protecting the community environment from the ravages of growth is quite the same. The best minds don't know for sure the best way to do it. There are few public establishments whose sole role is to preserve the environment. One can postpone assaults on the environment, but by and large, it takes eternal vigilance of concerned citizens, who, at best, can only reduce the rate of loss of the environment. There is no terminal time at which one can have a party to celebrate the fact that "We have saved the environment!"

Laws Relating to Sustainability

Let us be specific and state that both 'carrying capacity' and 'sustainable' imply 'for the period in which we hope humans will inhabit the earth'. This means 'for many millennia'.

Many prominent individuals have given postulates and laws relating to population growth and sustainability.

The Two 'Postulata' of Thomas Malthus

The reverend Thomas Malthus used these two assumptions as the basis of his famous essay two hundred years ago (Appleman 1976):

– First, that food is necessary to the existence of man;

– Secondly, that the passion between the sexes is necessary and will remain nearly in its present state.

Boulding's Three Theorems

These theorems are from the work of the eminent economist Kenneth Boulding (1971):

– First Theorem or 'The Dismal Theorem': If the only ultimate check on the growth of population is misery, then the population will grow until it is miserable enough to stop its growth;

– Second Theorem or 'The Utterly Dismal Theorem': This theorem states that any technical improvement can only relieve misery for a while, for so long as misery is the only check on population, the [technical] improvement will enable population to grow, and will soon enable more people to live in misery than before. The final result of [technical] improvements, therefore, is to increase the equilibrium population which is to increase the total sum of human misery;

– Third Theorem or 'The moderately cheerful form of the Dismal Theorem': Fortunately, it is not too difficult to restate the Dismal Theorem in a moderately cheerful form, which states that if something else, other than misery and starvation, can be found which will keep a prosperous population in check, the population does not have to grow until it is miserable and starves, and it can be stably prosperous.

Boulding continues: Until we know more, the Cheerful Theorem remains a question mark. Misery we know will do the trick. This is the only sure-fire automatic method of bringing population to an equilibrium. Other things may do it.

In another context, Boulding (1971 p. 361) observed,

"The economic analysis I presented earlier indicates that the major priority, and one in which the United Nations can be of great utility, is a world campaign for the reduction of birth rates. This, I suggest, is more important than any program of foreign aid and investments. Indeed, if it is neglected, all programs of aid and investment will, I believe, be ultimately self-defeating and will simply increase the amount of human misery."

Laws of Sustainability

The Laws that follow are offered to define the term 'sustainability'. In some cases these statements are accompanied by corollaries that are identified by capital letters. They all apply for populations and rates of consumption of goods and resources of the sizes and scales found in the world in 2005, and may not be applicable for small numbers of people or to groups in primitive tribal situations.

These Laws are believed to hold rigorously.

The list is but a single compilation, and hence may be incomplete. Readers are invited to communicate with the author in regard to items that should or should not be in this list:

First Law: Population growth and/or growth in the rates of consumption of resources cannot be sustained.

A) A population growth rate less than or equal to zero and declining rates of consumption of resources are a necessary, but not a sufficient, condition for a sustainable society.

B) Unsustainability will be the certain result of any program of 'development', that does not plan the achievement of zero (or a period of negative) growth of populations and of rates of consumption of resources. This is true even if the program is said to be 'sustainable'.

C) The research and regulation programs of governmental agencies that are charged with protecting the environment and promoting 'sustainability' are, in the long run, irrelevant, unless these programs address vigorously and quantitatively the concept of carrying capacities and unless the programs study in depth the demographic causes and consequences of environmental problems.

D) Societies, or sectors of a society, that depend on population growth or growth in their rates of consumption of resources, are unsustainable.

E) Persons who advocate population growth and/or growth in the rates of consumption of resources are advocating unsustainability.

F) Persons who suggest that sustainability can be achieved without stopping population growth are misleading themselves and others.

G) Persons whose actions directly or indirectly cause increases in population or in the rates of consumption of resources are moving society away from sustainability.

H) The term 'sustainable growth' is an oxymoron.

I) In terms of population sizes and rates of resource consumption, "The only smart growth is no growth." (Kerr 2002)

Second Law: In a society with a growing population and/or growing rates of consumption of resources, the larger the population, and/or the larger the rates of consumption of resources, the more difficult it will be to transform the society to the condition of sustainability.

Third Law: The response time of populations to changes in the human fertility rate is the average length of a human life, or approximately 70 years. (Bartlett and Lytwak 1995) [This is called 'population momentum'.]

A) A nation can achieve zero population growth if,

– the fertility rate is maintained at the replacement level for 70 years, and

– there is no net migration during the 70 years.

During the 70 years the population continues to grow, but at declining rates until the growth finally stops after approximately 70 years.

B) If we want to make changes in the total fertility rates so as to stabilize the population by the mid to late 21st century, we must make the necessary changes now.

C) The time horizon of political leaders is of the order of two to eight years.

D) It will be difficult to convince political leaders to act now to change course, when the full results of the change may not become apparent in the lifetimes of those leaders.

Fourth Law: The size of population that can be sustained (the carrying capacity) and the sustainable average standard of living of the population are inversely related to one another. (This must be true even though Cohen (1995) asserts that the numerical size of the carrying capacity of the Earth cannot be determined.)

A) The higher the standard of living one wishes to sustain, the more urgent it is to stop population growth.

B) Reductions in the rates of consumption of resources and reductions in the rates of production of pollution can shift the carrying capacity in the direction of sustaining a larger population.

Fifth Law: One cannot sustain a world in which some regions have high standards of living while others have low standards of living.

Sixth Law: All countries cannot simultaneously be net importers of carrying capacity.

A) World trade involves the exportation and importation of carrying capacity.

Seventh Law: A society that has to import people to do its daily work ("We can't find locals who will do the work") is not sustainable.

Eighth Law: Sustainability requires that the size of the population be less than or equal to the carrying capacity of the ecosystem for the desired standard of living.

A) Sustainability requires an equilibrium between human society and dynamic but stable ecosystems.

B) Destruction of ecosystems tends to reduce the carrying capacity and/or the sustainable standard of living.

C) The rate of destruction of ecosystems increases as the rate of growth of the population increases.

D) Affluent countries, through world trade, destroy the ecosystems of less developed countries.

E) Population growth rates less than or equal to zero are necessary, but are not sufficient, conditions for halting the destruction of the environment. This is true locally and globally.

Ninth Law (the lesson of *The Tragedy of the Commons*): The benefits of population growth and of growth in the rates of consumption of resources accrue to a few; the costs of population growth and growth in the rates of consumption of resources are borne by all of society (Hardin 1968).

A) Individuals who benefit from growth will continue to exert strong pressures supporting and encouraging both population growth and growth in rates of consumption of resources.

B) The individuals who promote growth are motivated by the recognition that growth is good for them. In order to gain public support for their goals, they must convince people that population growth and growth in the rates of consumption of resources, are also good for society. [This is the Charles Wilson argument: if it is good for General Motors, it is good for the United States (Yates 1983).]

Tenth Law: Growth in the rate of consumption of a non-renewable resource, such as a fossil fuel, causes a dramatic decrease in the life expectancy of the resource.

A) In a world of growing rates of consumption of resources, it is seriously misleading to state the life expectancy of a non-renewable resource 'at present rates of consumption', i.e., with no growth. More relevant than the life expectancy of a resource is the expected date of the peak production of the resource, i.e., the peak of the Hubbert curve (Hubbert 1972).

B) It is intellectually dishonest to advocate growth in the rate of consumption of non-renewable resources while, at the same time, reassuring people about how long the resources will last 'at present rates of consumption'. [zero growth]

Eleventh Law: The time of expiration of non-renewable resources can be postponed, possibly for a very long time, by:

– Technological improvements in the efficiency with which the resources are recovered and used;

– Using the resources in accord with a program of 'sustained availability'[3] (Bartlett 1986);

– Recycling;

– The use of substitute resources.

Twelfth Law: When large efforts are made to improve the efficiency with which resources are used, the resulting savings are easily and completely wiped out by the added resources that are consumed as a consequence of modest increases in population.

A) When the efficiency of resource use is increased, the consequence often is that the 'saved' resources are not put aside for the use of future generations, but instead are used immediately to encourage and support larger populations.

B) Humans have an enormous compulsion to find an immediate use for all available resources.

Thirteenth Law: The benefits of large efforts to preserve the environment are easily canceled by the added demands on the environment that result from small increases in human population.

Fourteenth Law (Second Law of Thermodynamics): When rates of pollution exceed the natural cleansing capacity of the environment, it is easier to pollute than it is to clean up the environment.

Fifteenth Law (Eric Sevareid's Law): The chief cause of problems is solutions (Sevareid 1970).

A) This law should be a central part of higher education, especially in engineering.

Sixteenth Law: Humans will always be dependent on agriculture. [This is the first of Malthus' two postulata.]

A) Supermarkets alone are not sufficient.

B) The central task in sustainable agriculture is to preserve agricultural land.

Agricultural land must be protected from losses due to things such as:

– Urbanization and development;

– Erosion;

– Poisoning by chemicals.

**Seventeenth Law: If, for whatever reason, humans fail to stop popula-
tion growth and growth in the rates of consumption of resources,
Nature will stop these growths.**

A) By contemporary Western standards, Nature's method of stopping
growth is cruel and inhumane.

B) Glimpses of Nature's method of dealing with populations that have
exceeded the carrying capacity of their lands can be seen each night on the
television news reports from places where large populations are experiencing starvation and misery.

**Eighteenth Law: In local situations within the US, creating jobs in-
creases the number of people locally who are out of work.**

A) Newly created jobs in a community temporarily lowers the unemployment rate (say from 5% to 4%), but then people move into the community to restore the unemployment rate to its earlier higher value (of 5%),
but this is 5% of the larger population, so more individuals are out of work
than before.

Nineteenth Law: Starving people don't care about sustainability.

A) If sustainability is to be achieved, the necessary leadership and resources must be supplied by people who are not starving.

**Twentieth Law: The addition of the word 'sustainable' to our vocabu-
lary, to our reports, programs, and papers, to the names of our
academic institutes and research programs, and to our community initiatives, is not sufficient to ensure that our society becomes sustainable.**

Twenty-First Law: Extinction is forever.

Where Do We Go from Here?

The challenge of making the transition to a sustainable society is enormous, in part because of a major global effort to keep people from recognizing the centrality of population growth to the enormous problems of the
US and the world.

On the global scale, we need to support family planning throughout the
world, and we should generally restrict our foreign aid to those countries
that make continued demonstrated progress in reducing population growth
rates and sizes.

The immediate task is to restore numeracy to the population programs in the local, national and global agendas.

On the national scale, we can work for the selection of leaders who will recognize that population growth is the major problem in the US and who will initiate a national dialog on the problem. With a lot of work at the grassroots, our system of representative government will respond.

On the local and national levels, we must focus serious attention and large fiscal resources on the development of renewable energy sources.

On the local and national levels, we need to work to improve social justice and equity.

On the community level in the US, we should work to make growth pay for itself.

Boulding on Malthus

In writing about Malthus' essay on population, Kenneth Boulding (1971) observed,

"... the essay, punctures the easy optimism of the utopians of any generation. But by revealing the nature of at least one dragon that must be slain before misery can be abolished, its ultimate message is one of hope, and the truth, however unpleasant, tends "not to create despair, but activity" of the right kind."

A Thought for the Future

When competing 'experts' recommend diametrically opposing paths of action regarding resources, carrying capacity, sustainability, and the future, we serve the cause of sustainability by choosing the conservative path, which is defined as the path that would leave society in the less precarious position in case the chosen path turns out to be the wrong path.

Endnotes

[1] This is a revised and shortened version of the paper that was previously published in Population & Environment, Vol. 16, No. 1, September 1994, pp. 5–35; in the Renewable Resources Journal, Vol. 15, No. 4, Winter 1997-98, pp. 6–23; in Focus, Vol. 9, No. 1, 1999, pp. 49–68; and as chapter 16 (with the additional title "The Great Challenge") in the anthology "Getting to the Source—Readings on Sustainable Values," edited by William Ross McCluney, SunPine Press, Cape Canaveral, Florida, 2004, pp. 165–205. With kind permission of Springer Science and Business Media.

[2] Willie Sutton was a legendary bank robber. When asked why he robbed banks, he is said to have responded, "That's where the money is!"

[3] 'Sustained availability' involves having the rate of use of a finite non-renewable resource decline steadily in a way that guarantees that the resource will last forever.

References

Abelson PH (1990) Dialog on the Future of Agriculture. *Science*, Vol. 249, p. 457

Appleman P ed (1976) *An Essay on the Principles of Population by Thomas Robert Malthus: Text, Sources, and Background Criticism.* WW Norton & Co, New York

Bartlett AA (1978) Forgotten Fundamentals of the Energy Crisis. *American Journal of Physics,* Vol 46, September 1978, pp. 876–888

Bartlett AA (1986) Sustained Availability: A Management Program for Non-Renewable Resources. *American Journal of Physics,* Vol. 54, pp. 398–402

Bartlett AA (1994) Reflections on Sustainability, Population Growth, and the Environment. *Population & Environment,* Vol. 16, No. 1, September 1994, pp. 5–35

Bartlett AA (1996) The Exponential Function, XI: The New Flat EarthSociety. *The Physics Teacher,* Vol. 34, September 1996, pp. 342–343

Bartlett AA (1997) Is There a Population Problem? *Wild Earth,* Vol. 7, No. 3, Fall 1997, pp. 88–90

Bartlett AA and Lytwak EP (1995) Zero Growth of the Population of the United States. *Population & Environment,* Vol. 16, No. 5, May 1995, pp. 415–428

Boulding K (1971) *Collected Papers, Vol II.* Foreword to T.R. Malthus, Population, The First Essay. Colorado Associated University Press, Boulder

Brundtland GH (1987) *Our Common Future.* World Commission on Environment and Development, Oxford University Press, Oxford New York

Clinton WJ (1992) *Putting People First: How We Can All Change America.* Times Books, New York

Cohen JE (1995) *How Many People Can the Earth Support?* WW Norton & Co, New York

Cole HSD, Freeman C, Jahoda M and Pavitt KLR eds (1973) *Models of Doom: A Critique of Limits to Growth.* Universe Books, New York

Daly HE (1994) *Sustainable Growth: An Impossibility Theorem.* Clearinghouse Bulletin, April 1994. Carrying Capacity Network, Washington DC

EPA: Environmental Protection Agency (1993) *Sustainable Development and the Environmental Protection Agency.* Report to the Congress, EPA 230-R-93-005, June 1993

Ehrlich PR and Ehrlich AH (1992) *The Most Overpopulated Nation.* The NPG Forum, Undated Monograph, Negative Population Growth, Teaneck NJ and Clearinghouse Bulletin, October 1992, Carrying Capacity Network, Washington DC, p. 1

Fodor E (1999) *Better not Bigger.* New Society Publishers, Gabriola Island BC, Canada

Giampietro M, Bukkens SGF and Pimentel D (1992) Limits to Population Size: Three Scenarios of Energy Interaction Between Human Society and Ecosystems. *Population and Environment,* Vol. 14, pp. 109–131

Hardin G (1968) The Tragedy of the Commons. *Science,* Vol. 162, pp. 1243–1248

Hubbert MK (1972) *U.S. Energy Resources: A Committee on Interior and Insular Affairs Report,* United States Senate, pursuant to Senate Resolution 45, National Fuels and Energy Policy Study, Serial No. 93-40 (92-75), Part 1, US Government Printing Office, Washington DC

Kendall HW and Pimentel D (1994) Constraints on the Expansion of the Global Food Supply. *Ambio,* Vol. 23, No. 3, May 1994, pp. 198–205

Kerr A (2002) *Endless Growth or The End of Growth?,* www.andykerr.net/Growth/EndlessGrowth.html, retrieved on July 18, 2005

May RM (1993) The End of Biological History? *Scientific American,* March 1993, pp. 146–149

Meadows DH, Meadows DL, Randers J and Behrens WW (1972) *Limits to Growth: A Report for the Club of Rome's Project on the Predicament of Mankind.* Universe Books, New York

Sevareid E (1970) CBS News, 29 December 1970. Quoted in Martin TL (1973) *Malice in Blunderland.* McGraw-Hill Book Co, New York

Simon J (1995) The State of Humanity: Steadily Improving. *Cato Policy Report,* September / October 1995, Vol. 17, No. 5

Wattenberg BJ (1997) *Boulder Daily Camera,* 30 November 1997. Reprinted from New York Times Magazine, 23 Nov 1997

Yates B (1983) The Decline and Fall of the American Automobile Industry. Empire Books, New York

Sustainable Development—
Definitions, Principles, Policies[1]

Herman E. Daly

Introduction

I begin by considering two competing definitions of sustainability, utilitybased versus throughput-based, and offer reasons for rejecting the former and accepting the latter. Next, I consider the concept of development as currently understood (GDP growth led by global economic integration) and why it conflicts with sustainability, as well as with the premises of comparative advantage. Then, I turn to the more general necessity of introducing the concept of throughput into economic theory, noting the awkward consequences to both micro and macro economics of having ignored the concept. Finally, I consider some policy implications for sustainable development that come from a more adequate economic theory. These policies (ecological tax reform and/or cap and trade limits on throughput) are based on the principle of frugality first, rather than efficiency first.

Definitions

Exactly what is it that is supposed to be sustained in 'sustainable' development? Two broad answers have been given:

First, utility should be sustained; that is, the utility of future generations is to be non-declining. The future should be at least as well off as the present in terms of its utility or happiness as experienced by itself. Utility here refers to average per capita utility of members of a generation.

Second, physical throughput should be sustained, that is, the entropic physical flow from nature's sources through the economy and back to nature's sinks, is to be nondeclining. More exactly, the capacity of the ecosystem to

39

M. Keiner (ed.), The Future of Sustainability, 39–53.
© 2006 Springer. Printed in the Netherlands.

sustain those flows is not to be run down. Natural capital is to be kept intact.[2] The future will be at least well off as the present in terms of its access to biophysical resources and services supplied by the ecosystem. Throughput here refers to total throughput flow for the community over some time period (i.e., the product of per capita throughput and population).

These are two totally different concepts of sustainability. Utility is a basic concept in standard economics. Throughput is not, in spite of the efforts of Kenneth Boulding and Nicholas Georgescu-Roegen to introduce it. So it is not surprising that the utility definition has been dominant. Natural capital is the capacity of the ecosystem to yield both a flow of natural resources and a flux of natural services. Keeping natural capital constant is often referred to as 'strong sustainability' in distinction to 'weak sustainability' in which the sum of natural and manmade capital is kept constant.

Nevertheless, I adopt the throughput definition and reject the utility definition, for two reasons. First, utility is non-measurable. Second, and more importantly, even if utility were measurable it is still not something that we can bequeath to the future. Utility is an experience, not a thing. We cannot bequeath utility or happiness to future generations. We can leave them things and, to a lesser degree, knowledge.[3] Whether future generations make themselves happy or miserable with these gifts is simply not under our control. To define sustainability as a non-declining intergenerational bequest of something that can neither be measured nor bequeathed strikes me as a nonstarter.[4] I hasten to add that I do not think economic theory can get along without the concept of utility. I just think that throughput is a better concept by which to define sustainability.

The throughput approach defines sustainability in terms of something much more measurable and transferable across generations—the capacity to generate an entropic throughput from and back to nature.[5] Moreover this throughput is the metabolic flow by which we live and produce. The economy in its physical dimensions is made up of things—populations of human bodies, livestock, machines, buildings, and artifacts. All these things are what physicists call 'dissipative structures' that are maintained against the forces of entropy by a throughput from the environment. An animal can only maintain its life and organizational structure by means of a metabolic flow through a digestive tract that connects to the environment at both ends. So too with all dissipative structures and their aggregate, the human economy.

Economists are very fond of the circular flow vision of the economy, inspired by the circulation of blood discovered by William Harvey (1628), emphasized by the Physiocrats, and reproduced in the first chapter of every economics textbook. Somehow the digestive tract has been less inspirational to economists than the circulatory system. An animal with a circulatory system, but no digestive tract, could it exist, would be a perpetual motion machine. Biologists do not believe in perpetual motion. Economists seem dedicated to keeping an open mind on the subject.

Bringing the concept of throughput into the foundations of economic theory does not reduce economics to physics, but it does force the recognition of the constraints of physical law on economics. Among other things, it forces the recognition that 'sustainable' cannot mean 'forever'.[6] Sustainability is a way of asserting the value of longevity and intergenerational justice, while recognizing mortality and finitude. Sustainable development is not a religion, although some seem to treat it as such. Since large parts of the throughput are nonrenewable resources the expected lifetime of our economy is much shorter than that of the universe. Sustainability in the sense of longevity requires increasing reliance on the renewable part of the throughput, and a willingness to share the nonrenewable part over many generations.[7] Of course longevity is no good unless life is enjoyable, so we must give the utility definition its due in providing a necessary baseline condition. That said, in what follows I adopt the throughput definition of sustainability, and will have nothing more to say about the utility definition.

Having defined 'sustainable' let us now tackle 'development'. Development might more fruitfully be defined as more utility per unit of throughput, and growth defined as more throughput. But since current economic theory lacks the concept of throughput, we tend to define development simply as growth in GDP, a value index that conflates the effects of changes in throughput and utility.[8] The hope that the growth increment will go largely to the poor, or at least trickle down, is frequently expressed as a further condition of development. Yet any serious policy of redistribution of GDP from rich to poor is rejected as 'class warfare' that is likely to slow GDP growth. Furthermore, any recomposition of GDP from private goods toward public goods (available to all, including the poor) is usually rejected as government interference in the free market—even though it is well known that the free market will not produce public goods. We are assured that a rising tide lifts all boats that the benefits of growth will eventually trickle down to the poor. The key to development is still aggregate growth, and the key to aggregate growth is currently thought to be global economic integration—free trade and free capital mobility. Export-led development is considered the only option. Import substitution is no longer mentioned, except to be immediately dismissed as 'discredited'.

Will this theory or ideology of "development as global growth" be successful? I doubt it, for two reasons, one having to do with environmental sustainability, the other with social equity:

– Ecological limits are rapidly converting economic growth into uneconomic growth—i.e., throughput growth that increases costs by more than it increases benefits, thus making us poorer not richer. The macroeconomy is not the Whole—it is Part of a larger Whole, namely the ecosystem. As the macroeconomy grows in its physical dimensions (throughput), it does not grow into the infinite Void. It grows into and encroaches upon the finite ecosystem, thereby incurring an opportunity cost of preempted natural capital and services. These opportunity costs (depletion, pollution, sacrificed ecosystem services) can be, and often are, worth more than the extra production benefits of the throughput growth that caused them. We cannot be absolutely sure because we measure only the benefits, not the costs.[9] We do measure the regrettable defensive expenditures made necessary by the costs, but even those are added to GDP rather than subtracted.

– Even if growth entailed no environmental costs, part of what we mean by poverty and welfare is a function of relative rather than absolute income, that is, of social conditions of distributive inequality. Growth cannot possibly increase everyone's relative income. Insofar as poverty or welfare is a function of relative income, then growth becomes powerless to affect it.[10] This consideration is more relevant when the growth margin is devoted more to relative wants (as in rich countries) than when devoted more to absolute wants (as in poor countries). But if the policy for combating poverty is global growth then the futility and waste of growth dedicated to satisfying the relative wants of the rich cannot be ignored.

Am I saying that wealth has nothing to do with welfare, and that we should embrace poverty? Not at all! More wealth is surely better than less, up to a point. The issue is, does growth increase net wealth? How do we know that throughput growth, or even GDP growth, is not at the margin increasing 'illth' faster than wealth, making us poorer, not richer?[11] Illth accumulates as pollution at the output end of the throughput, and as depletion at the input end. Ignoring throughput in economic theory leads to treating depletion and pollution as 'surprising' external costs, if recognized at all. Building the throughput into economic theory as a basic concept allows us to see that illth is necessarily generated along with wealth. When a growing throughput generates illth faster than wealth then its growth has become uneconomic. Since macroeconomics lacks the concept of throughput it is to be expected that the concept of 'uneconomic growth' will not make sense to macroeconomists.

While growth in rich countries might be uneconomic, growth in poor countries where GDP consists largely of food, clothing, and shelter, is still very likely to be economic. Food, clothing, and shelter are absolute needs, not self-canceling relative wants for which growth yields no welfare. There is much truth in this, even though poor countries too are quite capable of deluding themselves by counting natural capital consumption (depleting mines, wells, forests, fisheries, and topsoil) as if it were Hicksian income.[12] One might legitimately argue for limiting growth in wealthy countries (where it is becoming uneconomic) in order to concentrate resources on growth in poor countries (where it is still economic).

The current policy of the IMF, WTO and WB, however, is decidedly not for the rich to decrease their uneconomic growth to make room for the poor to increase their economic growth. The concept of uneconomic growth remains unrecognized. Rather the vision of globalization requires the rich to grow rapidly in order to provide markets in which the poor can sell their exports. It is thought that the only option poor countries have is to export to the rich, and to do that they have to accept foreign investment from corporations who know how to produce the high-quality stuff that the rich want. The resulting necessity of repaying these foreign loans reinforces the need to orient the economy towards exporting, and exposes the borrowing countries to the uncertainties of volatile international capital flows, exchange rate fluctuations, and unrepayable debts, as well as to the rigors of competing with powerful world-class firms.

The whole global economy must grow for this policy to work, because unless the rich countries grow rapidly they will not have the surplus to invest in poor countries, nor the extra income with which to buy the exports of the poor countries.

The inability of macroeconomists to conceive of uneconomic growth is very strange, given that microeconomics is about little else than finding the optimal extent of each micro activity. An optimum, by definition, is a point beyond which further growth is uneconomic. The cardinal rule of microeconomic optimization is to grow only to the point at which marginal cost equals marginal benefit. That has been aptly called the 'when to stop' rule—when to stop growing, that is. Macroeconomics has no 'when to stop' rule. GDP is supposed to grow forever.[13] The reason is that the growth of the macroeconomy is not thought to encroach on anything and thereby incur any growthlimiting opportunity cost. By contrast the microeconomic parts grow into the rest of the macroeconomy by competing away resources from other microeconomic activities thereby incurring an opportunity cost. The macroeconomy, however, is thought to grow into the infinite Void, never encroaching on or displacing anything of value. The point to be emphasized is that the macroeconomy too is a Part of a larger finite Whole, namely the ecosystem. The optimal scale of the macroeconomy relative to its containing ecosystem is the critical issue to which macroeconomics has been blind. This blindness to the costs of growth in scale is largely a consequence of ignoring throughput, and has led to the problem of ecological unsustainability.

Growth by Global Integration:
Comparative and Absolute Advantage
and Related Confusions

Under the current ideology of export-led growth the last thing poor countries are supposed to do is to produce anything for themselves. Any talk of import substitution is nowadays met by trotting out the abused and misunderstood doctrine of comparative advantage. The logic of comparative advantage is unassailable, given its premises. Unfortunately one of its premises (as emphasized by Ricardo) is capital immobility between nations. When capital is mobile, as indeed it is, we enter the world of absolute advantage, where, to be sure, there are still global gains from specialization and trade. However, there is no longer any guarantee that each country will necessarily benefit from free trade as under comparative advantage. One way out of this difficulty would be to greatly restrict international capital mobility thereby making the world safe for comparative advantage.[14] The other way out would be to introduce international redistribution of the global gains from trade resulting from absolute advantage. Theoretically the gains from absolute advantage specialization would be even greater than under comparative advantage because we would have removed a constraint to the capitalists' profit maximization, namely the international immobility of capital. But absolute advantage has the political disadvantage that there is no longer any guarantee that free trade will mutually benefit all nations. Which solution does the IMF advocate—comparative advantage vouch-safed by capital immobility, or absolute advantage with redistribution of gains to compensate losers? Neither. They prefer to pretend that there is no contradiction, and call for both comparative advantage-based free trade, and free international capital mobility—as if free capital mobility were a logical extension of comparative advantage-based free trade instead of a negation of its premise. This is incoherent.

In an economically integrated world, one with free trade and free capital mobility, and increasingly free, or at least uncontrolled, migration, it is difficult to separate growth for poor countries from growth for rich countries, since national boundaries become economically meaningless. Only by adopting a more nation-based approach to development can we say that growth should continue in some countries but not in others. But the globalizing trio, the IMF, WTO, and WB cannot say this. They can only advocate continual global growth in GDP. The concept of uneconomic growth just does not compute in their vision of the world. Nor does their cosmopolitan ideology recognize the nation as a fundamental unit of community and policy, even though their founding charter defines the IMF and World Bank as a federation of nations.

Ignoring Throughput in Macroeconomics: GDP and Value Added

As noted, throughput and scale of the macroeconomy relative to the ecosystem are not familiar concepts in economics. Therefore let us return for a while to the familiar territory of GDP and value added, and approach the concept of throughput by this familiar path. Economists define GDP as the sum of all value added by labor and capital in the process of production.[15] Exactly what it is that value is being added to is a question to which little attention is given. Before considering it let us look at value added itself.

Value added is simultaneously created and distributed in the very process of production. Therefore, economists argue that there is no GDP 'pie' to be independently distributed according to ethical principles. As Kenneth Boulding put it, instead of a 'pie', there are only a lot of little 'tarts' consisting of the value added by different people or different countries, and mindlessly aggregated by statisticians into an abstract 'pie' that doesn't really exist as an undivided totality. If one wants to redistribute this imaginary 'pie' one should appeal to the generosity of those who baked larger 'tarts' to share with those who baked smaller 'tarts', not to some invidious notion of equal participation in a fictitious common inheritance.

I have considerable sympathy with this view, as far as it goes. But it leaves out something very important.

In our one-eyed focus on value added we economists have neglected the correlative category, 'that to which value is added', namely the throughput. 'Value added' by labor and capital has to be added to something, and the quality and quantity of that something is important. There is a real and important sense in which the original contribution of nature is indeed a 'pie', a pre-existing, systemic totality that we all share as an inheritance. It is not an aggregation of little tarts that we each baked ourselves. Rather it is the seed, soil, sunlight, and rain from which the wheat and apples grew that we converted into tarts by our labor and capital. The claim for equal access to nature's bequest is not the invidious coveting of what our neighbor produced by her own labor and abstinence. The focus of our demands for income to redistribute to the poor, therefore, should be on the value of the contribution of nature, the original value of the throughput to which further value is added by labor and capital—or, if you like, the value of low entropy added by natural processes to neutral, random, elemental stuff.

Ignoring Throughput in Microeconomics: The Production Function

But there is also a flaw in our very understanding of production as a physical process. Neoclassical production functions are at least consistent with the national accountant's definition of GDP as the sum of value added by labor and capital, because they usually depict output as a function of only two inputs, labor and capital. In other words, value added by labor and capital in production is added to nothing, not even valueless neutral stuff. But value cannot be added to nothing. Neither can it be added to ashes, dust, rust, and the dissipated heat energy in the oceans and atmosphere. The lower the entropy of the input the more capable it is of receiving the imprint of value added by labor and capital. High entropy resists the addition of value. Since human action cannot produce low entropy in net terms we are entirely dependent on nature for this ultimate resource by which we live and produce (Georgescu-Roegen 1971). Any theory of production that ignores this fundamental dependence on throughput is bound to be seriously misleading.

As an example of how students are systematically misled on this issue I cite a textbook used in the microeconomic theory course at my institution. On p 146 the student is introduced to the concept of production as the conversion of inputs into outputs via a production function. The inputs or factors are listed as capital *(K)*, labor *(L)*, and materials *(M)*—the inclusion of materials is an unusual and promising feature (Perloff 2001). We turn the page to p.147 where we now find the production function written symbolically as $q = f(K, L)$. *M* has disappeared, never to be seen again in the rest of the book. Yet the output referred to in the text's 'real world example' of the production process is 'wrapped candy bars'. Where in the production function are the candy and wrapping paper as inputs?[16] Production functions are often usefully described as technical recipes. But unlike real recipes in real cookbooks we are seldom given a list of ingredients!

And even when neoclassicals do include resources as a generic ingredient it is simply R raised to an exponent and multiplied by L and K, also each raised to an exponent. Such a multiplicative form means that R can approach zero if only K and L increase sufficiently. Presumably we could produce a 100-pound cake with only a pound of sugar, flour, eggs, etc., if only we had enough cooks stirring hard in big pans and baking in a big enough oven!

The problem is that the production process is not accurately described by the mathematics of multiplication. Nothing in the production process is analogous to multiplication.[17] What is going on is transformation, a fact that is hard to recognize if throughput is absent. R is that which is being transformed from raw material to finished product and waste (the latter symptomatically is not listed as an output in production functions). R is a flow. K and L are agents of transformation, stocks (or funds) that effect the transformation of input R into output Q, but which are not themselves physically embodied in Q. There can be substitution between K and L, both agents of transformation, and there can be substitution among parts of R (aluminum for copper), both things undergoing transformation. But the relation between agent of transformation (efficient cause) and the material undergoing transformation (material cause), is fundamentally one of complementarity. Efficient cause is far more a complement than a substitute for material cause! This kind of substitution is limited to using a little extra labor or capital to reduce waste of materials in process—a small margin soon exhausted.[18]

Language misleads us into thinking of the production process as multiplicative, since we habitually speak of output as 'product' and of inputs as 'factors'. What could be more natural than to think that we multiply the factors to get the product! That, however, is mathematics, not production! If we recognized the concept of throughput we would speak of 'transformation functions', not production functions.

Opposite Problems: Non-Enclosure of the Scarce and Enclosure of the Non-Scarce

Economists have traditionally considered nature to be infinite relative to the economy, and consequently not scarce, and therefore properly priced at zero. But nature is scarce, and becoming more so every day as a result of throughput growth. Efficiency demands that nature's services be priced, as even Soviet central planners eventually discovered. But to whom should this price be paid? From the point of view of efficiency it does not matter who receives the price, as long as it is charged to the users. But from the point of view of equity it matters a great deal who receives the price for nature's increasingly scarce services. Such payment is the ideal source of funds with which to fight poverty and finance public goods.

Value added belongs to whoever added it. But the original value of that to which further value is added by labor and capital should belong to everyone. Scarcity rents to natural services, nature's value added, should be the focus of redistributive efforts. Rent is by definition a payment in excess of necessary supply price, and from the point of view of market efficiency is the least distorting source of public revenue.

Appeals to the generosity of those who have added much value by their labor and capital are more legitimate as private charity than as a foundation for fairness in public policy. Taxation of value added by labor and capital is certainly legitimate. But it is both more legitimate and less necessary after we have, as much as possible, captured natural resource rents for public revenue.

The above reasoning reflects the basic insight of Henry George, extending it from land to natural resources in general. Neoclassical economists have greatly obfuscated this simple insight by their refusal to recognize the productive contribution of nature in providing 'that to which value is added'. In their defense it could be argued that this was so because in the past economists considered nature to be non-scarce, but now they are beginning to reckon the scarcity of nature and enclose it in the market. Let us be glad of this, and encourage it further.

Although the main problem I am discussing is the non-enclosure of the scarce, an opposite problem (enclosure of the non-scarce) should also be noted. There are some goods that are by nature non-scarce and non-rival, and should be freed from illegitimate enclosure by the price system. I refer especially to knowledge. Knowledge, unlike throughput, is not divided in the sharing, but multiplied. There is no opportunity cost to me from sharing knowledge with you. Yes, I would lose the monopoly on my knowledge by sharing it, but we economists have long argued that monopoly is a bad thing because it creates artificial scarcity that is both inefficient and unjust. Once knowledge exists, the opportunity cost of sharing it is zero and its allocative price should be zero. Consequently, I would urge that international development aid should more and more take the form of freely and actively shared knowledge, and less and less the form of interest-bearing loans. Sharing knowledge costs little, does not create unrepayable debts, and it increases the productivity of the truly scarce factors of production.

Although the proper allocative price of existing knowledge is zero, the cost of production of new knowledge is often greater than zero, sometimes much greater. This of course is the usual justification for intellectual property rights in the form of patent monopolies. Yet the main input to the production of new knowledge is existing knowledge, and keeping the latter artificially expensive will certainly slow down production of the former. This is an area needing much reconsideration. I only mention it here, and signal my skepticism of the usual arguments for patent monopolies, so emphasized recently by the free-trading globalizers under the gratuitous rubric of 'traderelated intellectual property rights'. As far as I know, James Watson and Francis Crick receive no patent royalties for having unraveled the structure of DNA, arguably the most basic scientific discovery of the twentieth century. Yet people who are tweaking that monumental discovery are getting rich from monopolizing their relatively trivial contributions that could never have been made without the free knowledge supplied by Watson and Crick.

Although the main thrust of my remarks is to bring newly scarce and truly rival natural capital and services into the market enclosure, we should not overlook the opposite problem, namely, freeing truly non-rival goods from their artificial enclosure by the market.

Principles and Policies for Sustainable Development

I am not advocating revolutionary expropriation of all private property in land and resources. If we could start from a blank slate I would be tempted to keep land and minerals as public property. But for many environmental goods, previously free but increasingly scarce, we still do have a blank slate as far as ownership is concerned. We must bring increasingly scarce yet unowned environmental services under the discipline of the price system, because these are truly rival goods the use of which by one person imposes opportunity costs on others.[19] But for efficiency it matters only that a price be charged for the resource, not who gets the price. The necessary price or scarcity rent that we collect on newly scarce environmental public goods (e.g., atmospheric absorption capacity, the electromagnetic spectrum) should be used to alleviate poverty and finance the provision of other public goods.

The modern form of the Georgist insight is to tax the resources and services of nature (those scarce things left out of both the production function and GDP accounts)—and to use these funds for fighting poverty and for financing public goods. Or we could simply disburse to the general public the earnings from a trust fund created by these rents, as in the Alaska Permanent Fund, which is perhaps the best existing institutionalization of the Georgist principle. Taking away by taxation the value added by individuals from applying their own labor and capital creates resentment. Taxing away value that no one added, scarcity rents on nature's contribution, does not create resentment. In fact, failing to tax away the scarcity rents to nature and letting them accrue as unearned income to favored individuals has long been a primary source of resentment and social conflict.

Charging scarcity rents on the throughput of natural resources and redistributing these rents to public uses can be effected either by ecological tax reform (shifting the tax base away from value added and on to throughput), or by quantitative cap-and-trade systems initiated by a government auction of pollution or depletion quotas. In differing ways each would limit throughput and expansion of the scale of the economy into the ecosystem, and also provide public revenue. I will not discuss their relative merits, having to do with price versus quantity interventions in the market, but rather emphasize the advantage that both have over the currently favored strategy. The currently favored strategy might be called 'efficiency first' in distinction to the 'frugality first' principle embodied in both of the throughput-limiting mechanisms mentioned above.[20]

'Efficiency first' sounds good, especially when referred to as 'win-win' strategies or more picturesquely as 'picking the low-hanging fruit'. But the problem of 'efficiency first' is with what comes second. An improvement in efficiency by itself is equivalent to having a larger supply of the factor whose efficiency increased. The price of that factor will decline. More uses for the now cheaper factor will be found. We will end up consuming more of the resource than before, albeit more efficiently. Scale continues to grow. This is sometimes called the 'Jevons effect'. A policy of 'frugality first', however, induces efficiency as a secondary consequence; 'efficiency first' does not induce frugality—it makes frugality less necessary, nor does it give rise to a scarcity rent that can be captured and redistributed.

I am afraid I will be told by some of my neoclassical colleagues that frugality is a value-laden concept, especially if you connect it with redistribution of scarcity rents to the poor. Who am I, they will ask, to impose my personal elitist preferences on the democratic marketplace, blah, blah, etc. etc. I am sure everyone has heard that speech. The answer to such sophistry is that ecological sustainability and social justice are fundamental objective values, not subjective individual preferences. There really is a difference, and it is past time for economists to recognize it.

Conclusion

Reducing poverty is indeed the basic goal of development, as the World Bank now commendably proclaims. But it cannot be attained by growth for two reasons. First, because growth in GDP has begun to increase environmental and social costs faster than it increases production benefits. Such uneconomic growth makes us poorer, not richer. Second, because even truly economic growth cannot increase welfare once we are, at the margin, producing goods and services that satisfy mainly relative rather than absolute wants. If welfare is mainly a function of relative income then aggregate growth is selfcanceling in its effect on welfare. The obvious solution of restraining uneconomic growth for rich countries to give opportunity for further economic growth, at least temporarily, in poor countries, is ruled out by the ideology of globalization, which can only advocate global growth. We need to promote national and international policies that charge adequately for resource rents, in order to limit the scale of the macroeconomy relative to the ecosystem and to provide a revenue for public purposes. These policies must be grounded in an economic theory that includes throughput among its most basic concepts. These efficient national policies need protection from the cost-externalizing, standardslowering competition that is driving globalization. Protecting efficient national policies is not the same as protecting inefficient national industries.

Endnotes

[1] This chapter is a revised version of the invited address "Sustainable Development: Definitions, Principles, Policies" at the World Bank, April 30, 2002, Washington DC
[2] Natural capital is the capacity of the ecosystem to yield both a flow of natural resources and a flux of natural services.
[3] To a lesser degree because knowledge must be actively learned anew each generation. It cannot simply be passively inherited.
[4] It also puts the future at a disadvantage—the present could bequeath an ever smaller throughput, and claim that this is sufficient for non declining utility if only the future takes full advantage of foreseeable possibilities of substitution in both production and utility functions. But if these substitution possibilities are so easy to foresee, then let the present take advantage of them now and thereby reduce its utility cost of a given throughput bequest.
[5] The throughput is not only measurable in principle but has been measured for several industrial countries in the pioneering physical accounting studies

published by WRI in collaboration with Dutch, German, Japanese, and Austrian research institutes. See Adriaanse et al. (1997), and Matthews et al. (2000).

[6] Science tells us the physical world will end either in the big cooling or the big crunch. 'Forever' requires a 'new creation'—death and rebirth, not perpetual extension. Economics is not eschatology.

[7] Investing non-renewable resource rents in renewable substitutes is a good policy, with impeccable neoclassical roots, for sustaining the throughput over a longer time.

[8] The prices used in calculating this value index are of course affected by the distributions of wealth and income, as well as by the exclusion of the demand of future generations and non human species, and by the failure to have included other external costs and benefits into prices. It is hard to give a normative meaning to an index constructed with such distorted relative prices.

[9] Evidence that growth in the US since the 1970s has likely been uneconomic is presented in Daly and Cobb (1994) appendix on the Index of Sustainable Economic Welfare.

[10] If welfare is a function of relative income, and growth increases everyone's income proportionally, then no one is better off. If growth increases only some incomes, then the welfare gains of the relatively better off are cancelled by the losses of the relatively worse off.

[11] 'Illth' is John Ruskin's useful term for the opposite of wealth, i.e., an accumulated stock of bads as opposed to a stock of goods.

[12] Instead of 'deluding themselves' perhaps I should say 'being deluded' by IMF and World Bank economists who require this misleading system of national accounts of them.

[13] Macroeconomists do recognize that the economy can grow too fast when it causes inflation, even though the economy can never be too big in their view.

[14] How might capital flows be restricted? A Tobin tax; a minimum residence time before foreign investment could be repatriated; and most of all something like Keynes' International Clearing Union in which multilateral balance on trade account is encouraged by charging interest on both surplus and deficit balances on current account. To the extent that current accounts are balanced, then capital mobility is correspondingly restricted.

[15] Note that GDP does not value resources (that to which value is added). Yet we all pay a price in the market for gasoline. That gasoline price, however, reflects the labor and capital expended in drilling, pumping, and refining the petroleum, not the value of petroleum *in situ,* which is taken as zero. Your uncle in Texas discovered oil on his ranch and Texaco is paying him for the right to extract it. Is that not a positive price for petroleum *in situ*? It looks like it, but the amount Texaco will pay your uncle is determined by how easy it is to extract his oil relative to marginal deposits. Thus it is labor and capital saved in extraction that determines the rent to your uncle, not the value of oil *in situ* itself, which is still counted as zero.

[16] Some readers may rush to the defense of the textbook and tell me that the production function is only describing value added by L and K and that is why they

omitted material inputs. Let me remind such readers that on the previous page they included material inputs, and further that the production function is in units of physical quantities, not values or value added. Even if expressed in aggregate units of 'dollar's worth', it remains the case that a 'dollar's worth' of something is a physical quantity.

[17] I should say that I am thinking of the unit process of production—one laborer with one saw and one hammer converts lumber and nails into one doghouse in one period of time. We could of course multiply the unit process by ten and get ten doghouses made by ten laborers, etc. My point is that the unit process of production, which is what the production function describes, involves no multiplication.

[18] Of course one might imagine entirely novel technologies that use totally different resources to provide the same service. This would be a different production function, not substitution of factors within a production function. And if one wants to induce the discovery of new production functions that use the resource base more efficiently, then it would be a good idea to count resources as a factor of production in the first place, and to see to it that adequate prices are charged for their use! Otherwise such new technologies will not be profitable.

[19] For example, rents can be collected on atmospheric sink capacity, electromagnetic broadcast spectrum, fisheries, public timber and pasture lands, offshore oil, rights of way, orbits, etc.

[20] By 'frugality' I mean 'non-wasteful sufficiency', rather than 'meager scantiness'.

References

Adriaanse A, Bringezu S, Hammond A, Moriguchi Y, Rodenburg E, Rogich D, and Schütz H (1997) *Resource Flows: The Material Basis of Industrial Economies.* World Resources Institute, Washington DC

Daly H (1996) *Beyond Growth.* Beacon Press, Boston MA

Daly H and Cobb J (1994) *For the Common Good.* Beacon Press, Boston MA

Georgescu-Roegen N (1971) *The Entropy Law and the Economic Process.* Harvard University Press, Cambridge MA

Matthews E, Amann C, Fischer-Kowalski M, Bringezu S, Hüttler W, Kleijn R, Moriguchi Y, Ottke C, Rodenburg E, Rogich D, Schandl H, Schütz H, van der Voet E, and Weisz H (2000) *The Weight of Nations: Material Outflows from Industrial Economies.* World Resources Institute, Washington DC

Perloff JM (2001) *Microeconomics* (second edition). Addison Wesley, New York

Sustainability is Not Enough

PETER MARCUSE

"To think that their present circumstances and their present societal ar-
rangements might be sustained—that is an unsustainable thought for the
majority of the world's people."[1]

Programs and policies can be sustainable and socially just but, unfortunately, they can also be sustainable and unjust. On the other hand, unsustainable programs may be very just but fortunately some very unjust programs
are also unsustainable. Examples are easy: social security for the aged has
proven to be both socially desirable and very sustainable; but free reign
and legal protection for real estate speculation are, in the opinion of most
urbanists, very detrimental to a socially desirable environment although
they seem to be quite sustainable at present. On the other hand, publicly financed, owned and operated public housing is seen by many as very desirable but also appears unsustainable on any large scale in most countries;
also, forcible evictions without due process of law seems a more and more
unsustainable practice in most countries. Sustainability and social justice
do not necessarily go hand in hand. Sustainable, at least in its literal meaning "capable of being upheld or defended,"[2] requires careful examination if
we are to use it meaningfully in the arena of housing and urban development policy.

In this chapter I want to make several points:

– Sustainability is not a goal for a program—many bad programs are sustainable—but a constraint; its absence may limit the usefulness of a
  good program;

– While sustainability may be a useful formulation of goals on environmental issues, it is a treacherous one for urban policy because it suggests the possibility of a conflict-free consensus on policies whereas, in
  fact, vital interests do conflict; it will take more than simply better
  knowledge and a clearer understanding to produce change;

M. Keiner (ed.), The Future of Sustainability, 55–68.
© 2006 *Springer. Printed in the Netherlands.*

- Even in the environmental arena, sustainability cannot be the sole criterion by which programs are judged except in the, not useful, very long
  term because environmental policies must also take into account considerations of, for example, social justice;
- If sustainability means the ability not only to formulate and operate a
  desirable urban program but also to see it continue without detracting
  from other, also desirable, goals, then the concept may usefully emphasize the importance of long-term practicality to the consideration of such
  programs.

Sustainability is both an honorable goal for carefully defined purposes and
a camouflaged trap for the well-intentioned unwary. As a concept and a
slogan it has an honorable pedigree[3] in the environmental movement,
which has, by and large, succeeded in its fight to have the standard of sustainability generally accepted by all sides, at least in principle, although, in
practice, severe conflicts of interest still beset efforts to establish specific
standards. Few, these days, would contest that sustainability is something
desirable in environmental terms and that represents a substantial victory
for the environmental cause.

But the situation is quite different when it comes to other causes where,
I will contend, sustainability is not an appropriate goal; at best it is one criterion among others, not a goal. Its acceptance would not constitute an
achievement in the cause of better housing or better cities. The acceptance
of sustainability, at least in principle, in the environmental arena by virtually all actors[4] has led to the desire to use such a universally acceptable
goal as a slogan also in campaigns that have nothing to do with the environment but where the lure of universal acceptance is a powerful attraction.
Yet, in these other areas—and I focus on housing and urban development
as examples—'sustainability' is a trap. It suggests all humanity has a similar interest in 'sustainable housing' or 'sustainable urban development';
that if we all simply recognized our common interests everything would be
fine, we could end poverty, exploitation, segregation, inadequate housing,
congestion, ugliness, abandonment and homelessness. Yet, in these areas,
the idea of universal acceptance of meaningful goals is a chimera. Housing
and urban development are conflict-laden arenas: what benefits one hurts
another. A landlord's profits are at a tenant's expense; high-rise construction casts shadows on neighboring land uses; accessibility for one is pollution for another; security for some is taken to mean exclusion of others;
profit for business owners may mean layoffs for that business's workers.
Even ideologically, the parallel with environmental issues is deceptive. It
is hard to argue that a little short-term pollution contributes to a better

long-term environment but the argument is heard constantly that a few layoffs now will lead to increased competitiveness and fewer layoffs later.

I suggest, then, that 'sustainability' as a goal for housing or urban development just does not work.[5] In the first place, sustainability is not a goal; it is a constraint on the achievement of other goals.[6] Look at the early,[7] and still standard, definition of the World Commission on Environment and Development (the Brundtland Commission) in 1987:

> *"Sustainable development is development that meets the needs of the present without compromising the ability of future generations to meet their own needs."*[8]

Clearly, here, the goal is 'meeting the needs' and the remainder, 'making it sustainable', is obviously a constraint on the appropriate means to be used (WCED 1987).[9] Other formulations, defining sustainable development through a 'rule of constant capital' in which the goal is to pass on to the future the same stock of 'capital' as we have today, seem to drop the broad goal entirely and simply require that the human and natural capital (a perversion of the term?) of one generation be passed on unimpaired to the next. Others focus on the 'carrying-capacity of supporting ecosystems',[10] a much more questionable concept from the outset.[11]

No one who is interested in justice wants to sustain things as they are now. Sustainability plays very differently in the environmental sphere, where the whole point is simply that conditions as they are *cannot* be sustained and the only question is how rapidly to ameliorate them. If the environmental status quo was sustainable, environmentalists would be without a cause (Marcuse 1974). That perception is hardly prevalent in urban affairs or housing—we would hardly be satisfied if only present conditions could be sustained. In terms of our focus here, 'sustainability' taken as a goal by itself only benefits those who already have everything that they want. Indeed, even focusing on environmental concerns, the problem for most of the World's poor is not that their conditions *cannot* be sustained but that they *should not* be sustained.

Sustainability as a goal in itself, if we are to take the term's ordinary meaning, is the preservation of the status quo. It would, taken literally,[12] involve making only those changes that are required to maintain that status. Presumably, that is what the World Economic Forum, held in Davos, Switzerland in 1995, had in mind when it chose as its theme 'sustaining globalization'.[13] One might argue that the status quo is not sustainable *socially* because an unjust society will not endure. That is more a hope than a demonstrated fact. Indeed, the argument that the trouble with present urban conditions is that they are not sustainable opens the door to a fearsome debate of six decades ago in which the durability of some form of fascism was debated and indeed widely conceded on all sides. Unjust regimes have not always historically been the most short-lived ones. Teleological views of history are out of fashion and the 'end of history' argument is, rather, that the present is so sustainable that basic change is no longer conceivable, even if it were desirable.

Alternatively, one might argue, and with more evidence, that the status quo is not sustainable in strictly environmental terms; indeed, that is the origin of the 'sustainability' slogan.[14] But changes within the present system may be targeted at problems of environmental degradation, global warming, etc., while leaving other key undesirable aspects, such as social injustice, intact.[15] Presumably, good planning calls for social justice as well as environmental sustainability, not just the one or the other.

The more logically defensible use of the concept of sustainability might be to consider it as a constraint: any measure, desirable on other grounds, to meet substantive goals must *also* be capable of being maintained and contribute to the desired goal in the long run.[16] Here again, we run into problems if we are not careful to distinguish a constraint from a goal. If the sustainability of a measure is taken as a goal, the term can become either tautological or perverse. If a desired measure is socially just, the argument could go, then, and only then, is it sustainable?[17] (Any other argument would allow the conclusion that an unjust measure would be sustainable and, if that were so, would we want it or would we not reject the criterion of sustainability as validating it?) So, if justice is the standard by which sustainability is measured, why add the criterion of sustainability in judging the measure at all? Why not simply ask if it is just? Sustainability becomes tautological here. Presumably one does not want the perverse result that whatever can be kept up in the long run is good; the more effective the dictatorship, then, for instance, the better it would be.

If, however, sustainability is a constraint rather than a goal, then it can be used as a criterion to evaluate measures that achieve otherwise defined desirable goals; a desirable measure that is not sustainable is not as good as an equally desirable measure that is.[18] This goes beyond the Brundtland Commission definition, which simply requires no harm in the long run. It means that 'sustainability' is used to ask, in effect, what will be the long-term consequences of a given action or proposal? 'Sustainability' is not an independent goal, the contribution to which is to be weighed along with justice, etc. in evaluating a policy: A bad policy that is sustainable is not better than a bad policy that is unsustainable.[19] Sustainability is a limitation to be viewed in the context of an evaluation of the desirability, on substantive criteria, of other measures.[20] Balancing is required: A very good program that is not sustainable may be more desirable than a minor one that is. It may be more desirable to build 1,000 houses for low-income people this year, even if the pace cannot be sustained, rather than ten a year for the indefinite future.[21]

Perhaps 'sustainable' should only mean sustainable physically, environmentally, in the long run? That is a possible interpretation,[22] a modest one indeed, but perhaps a sustainable one? It would mean that our call for a sustainable living environment simply means focusing on the constraint of environmental sustainability? But even that limited use of 'sustainable' as 'environmentally sustainable' raises questions. For certainly, many desirable measures have an immediate adverse effect on the environment: building housing for low-income families on open land in a possible conservation area might be a classic example.[23] Or, the reverse situation, a short-term or limited measure protecting the environment may contribute to larger longer-term damage: saving electricity in a sprawling suburban development, for instance.[24] Indeed,

"There seems to be no place for cities in ecological design. If we look at each landscape separately, we are unable to ecologically justify plans for dense urban development. From a regional perspective, however, aggregation of urban and residential land uses may in fact be preferable."[25]

Two quite separate problems arise here, one social and political, the other scientific.

Socially, the costs of moving towards environmental sustainability (like the costs of environmental degradation)[26] will not be borne equally by everyone. In conventional economic terms, different people have different discount rates for the same cost or benefit. Meeting higher environmental standards increases costs. Some will profit from supplying the wherewithal to meet those standards; others, not being able to pay for them, will have to do without. The effects of income inequality are likely to be aggravated by such a raising of standards. We encounter the problem internationally in connection with issues such as atomic power plants in developing countries without other available sources of energy or in the rainforest disputes in South America. They are paralleled by other issues raised in the environmental justice movement in the United States. Better environments for some will be at the expense of worse environments for others, as waste disposal sites, air pollution and water contamination are moved around. Even when there is a solution that improves conditions for some without hurting others, the benefits will be unevenly distributed; costs and benefits to different groups and individuals cannot be simply netted out in quantitative terms.[27] The balancing act is often difficult indeed. What is clear is that the simple criterion of sustainability does not get us far.[28]

Indeed, the very definition of 'better environment' varies, in practice, by class and poverty level. As McGranahan, Sonsore and Kjellen (1996) point out, the issues tend to vary by scale and class. In the United States (and perhaps not only in the United States—certainly historically in South Africa, also I suspect increasingly in England and, to varying degrees, elsewhere) race plays a central role: The differential location of toxic waste sites by racial composition of surroundings is a classic example. For the poor, the issues tend to be immediate and very local: Water supply and waste disposal are immediate environmental problems. The affluent can escape these problems by choice of neighborhood or private market provision; their problems tend to be on a larger scale: automobile pollution at a city level, perhaps, global warming at a national or worldwide level. The agenda even for an environmentally limited definition of sustainability will be very different for different groups.

Scientifically, our knowledge is limited and the further into the future we wish to project it, the more the uncertainties grow. Malthus, who might uncharitably be called the grandfather (and the Club of Rome its father?) of the environmental sustainability movement, calculated with the best of the scientific knowledge of his day that food production would not sustain a world population much beyond its size at the time he wrote. Since then, it has increased more than five-fold, and is better nourished and lives longer. We know we need to deal with the problem of global warming and we know that relying on technological fixes is dangerous. Those two propositions should lead us to scale down certain activities linked to growth and to seek substitutes for others; they mandate the adoption of a limited set of specific policies to achieve specific goals by specific actors in a specific timetable. But, apart from those specific policies, a great deal is uncertain. Valid long-range concerns do not help very much in reaching a conclusion on even medium-range questions.

In any event, environmental long-term considerations are not the only ones that need to be taken into account when making decisions.[29] Other goals weigh in and other constraints need to be brought into the balance. Matters of social justice, of economic development, of international relations, of democracy, of democratic control over technological change and globalization also have both short and long-term implications. For a given policy to be desirable it must meet the constraints of sustainability in each of these dimensions; failure in any one is, in theory, sufficient cause for rejection. Environmental sustainability seems at first blush to be the most 'objective', the most inescapable, of all these constraints: if humankind dies off, the game is over. But may that not ultimately be said also if freedom or democracy or tolerance disappeared? Since none of any such events will be one-shot catastrophes, is the danger of environmental degradation greater today than that of war, fascism, poverty, hunger, disease or impoverishment for large numbers of people?

The problem of balancing differing goals and constraints is a well-recognized one. There is, for instance, an important debate on the relationship between growth and development (see i.e., Hamm and Muttagi eds 1992), a difficult issue and one viewed very differently in the developed as against the developing world. The discussion of sustainability has made a significant contribution to advancing the understanding of policy alternatives and their implications, but it is not quite clear why using the concept 'sustainable' in only half of the balancing equation clarifies the debate.

If we want to talk about sustainability as a constraint affecting all goals, we not only have to face the balancing problem: We also have to recognize the practical fact that sustainability in most usages is heavily focused on ecological concerns. That is not surprising, considering that 'sustainability' had its origins in the environmental movement. But why, given limited resources and limited power to bring about change, are efforts in the real world thus focused? What are the politics of the environmental sustainability movement? I would suggest that it is not for reasons of logic, not because the difficult issues of balance have been faced and brought to that conclusion but because of much more pragmatic concerns: that the environmental movement is a multi-class, if not indeed upper and middle-class, movement in its leadership, financing and political weight. While the environmental justice movement is making a substantial contribution both to social justice and to environmental protection, the environmental movement as a whole often proclaims itself to be above party, above controversy, seeking solutions from which everyone will benefit, to which no one can object. Thus, we get the report of a two-day workshop of the Sustainable Cities Program (1998) stating,

"One of the most important conclusions of the meeting was that implementation of concrete actions is often hampered by a variety of obstacles, and the meeting therefore recommended and agreed that the forthcoming annual meeting of the SCP be centered around this key theme."

How nice it would be if the next meeting figured out how to get over this variety of obstacles so that we could go on to other things! Perhaps it will build on the 'tool development activities' of the SCP and utilize its process.

The SCP process consists of a logically sequenced and interactive set of key activities whose systematic implementation and infusion into the existing institutions would bring profound changes in management approaches and improvement in information, decision making and implementation. The process forms the basis of the Source Book series (idem p. 3).

Maybe the next workshop will find a program we can all rally around and escape the unpleasant business of facing conflicting interests, having to deal with the unequal distribution of power, the necessities of redistribution, the defeats that accompany the victories? No wonder 'sustainability' is an attractive slogan, with such a hope! But if the goal is redistributing wealth or opportunity, or sharing power or reducing oppression, sustainability does not get us far.

To the extent that sustainability requires the review of policies designed today to meet the needs of today in such a way that they do not make things worse in the future, it is an important, if for planners not very new, concept. It might then be reformulated, to build on the words of the Brundtland Commission (WCED 1987):

"Sustainable development is development that meets specific needs of the present, and can be maintained into the future, without detracting from the satisfaction of other needs in the present or future."

It then amounts to little more than a call for long-term planning, something that has always been planners' bread and butter but adds perhaps a little more emphasis on long-term implications.

But the pursuit of sustainability is a snare and a delusion to the extent that calling for 'sustainable' activities in any sphere, albeit housing, planning, infrastructure, economic development, etc., suggests that there are policies that are of universal benefit, that everyone, every group, every interest will or should or must accept for their own best interests. If the appeal for sustainability implies that only our ignorance or stupidity prevents us from seeing what we all need, and prevents us from doing it,[30] it can undercut real reform. Indeed, a just, humane and environmentally sensitive world will, in the long run, be better for all of us. But getting to the long run entails conflict and controversy, issues of power and the redistribution of wealth. The frequent calls for 'us' to recognize 'our' responsibility for the environment avoids the real questions of responsibility, the real causes of pollution and degradation.[31] The slogan of 'sustainability' hides rather than reveals that unpleasant fact.

We should rescue sustainability as an honorable, indeed critically important, goal for environmental policy by confining its use only to where it is appropriate, recognizing its limitations and avoiding the temptation to take it over as an easy way out of facing the conflicts that beset us in other areas of policy. If we do feel called upon to use it in the area of social policy, it should be to emphasize the criterion of long-term political and social viability in the assessment of otherwise desirable programs and not as a goal replacing social justice, which must remain the focal point for our efforts.

Endnotes

[1] The formulation is a reworking of an aphorism of the Berlin Institute for Critical Theory which, building on Walter Benjamin's "The concept of progress should be grounded in the idea of catastrophe," adds: That things 'just keep on going' *is* the catastrophe." Inkrit, Conference Announcement, July 9, 1998.

[2] Oxford English Dictionary (1971) Compact Edition, p. 3,191. The etymology derives the word from *tenire*, "to hold," thus capable of being held on to.

[3] For a brief history of its current usage, see Voula Mega (1996) one of the leading researchers in the area. David Satterthwaite, of the International Institute for Environment and Development, has pointed out to me Barbara Ward's use of the phrase in the early 1970s, in very much the Brundtland Commission's sense, and its somewhat unthinking adoption as a catchword by many international development agencies to mean, simply, funded projects that could survive without falling apart in the medium to long-term. Letter dated July 6, 1998.

[4] David Satterthwaite comments on this phenomenon, and points out its potential as an escape from recognizing direct responsibilities, in an excellent article I saw subsequent to writing this paper: see Satterthwaite (1997).

[5] I have in mind formulations, such as, the goal is the "development of a housing system that is sustainable for people and the planet." See Bhatti et al. eds (1994).

[6] After this was written, I came across a discussion, which raised some similar issues as raised here: "... the primary environmental concerns of the more disadvantaged urban dwellers are not issues of sustainability, narrowly defined. Should a broader definition of sustainability be adopted or should the pre-eminence of sustainability concerns be rejected? ... Should the definition be reworked or ... sustainability ... be only one objective or constraint, among many?" See McGranahan et al. (1996). Without resolving the question as a theoretical one, the paper goes on to point out the differentiated views on the issue by class.

[7] The earliest formal usage I have found is in UNESCO's Man and the Biosphere Programme in the early 1970s, followed by explicit focus on the term in the World Conservation Strategy of the International Union for the Conservation of Nature, although it was strictly limited to environmental aspects. See Lawrence (1996).

[8] This and the following discussion draws on the European Foundation for the Improvement of Living and Working Conditions (1998) For an alternative formulation, see the suggestion at the conclusion of this paper.

[9] The same is true of William Rees' definition: "...positive socio-economic change that does not undermine the ecological and social systems upon which communities and societies are dependent" in William Rees (1988).

[10] The World Conservation Union, UNEP and WWF; see contributions to Price and Tsouris (1996).

[11] This does not apply, of course, to the environmental justice movement whose issue is the discriminatory impact of environmental degradation. The distribu-

tion of the costs and benefits of achieving a sustainable environment remain an issue even were the goal of sustainability to be achieved, but it then becomes an issue of justice, not of sustainability.

[12] On the other hand, its meaning can be made elastic and be redefined to encompass many other goals; but then the usefulness of the term evaporates. "A sustainable city is one which succeeds in balancing economic, environmental and socio-cultural progress through processes of active citizen participation" quoted in Mega and Pedersen (1997). Or take the even more far-reaching use in AHURI's 1997 catalogue of publications: "Sustainable issues ... are taken as a general umbrella term incorporating research into processes of urbanization, globalization and economic restructuring, their urban and regional impacts, urban metabolism as a framework for analyzing quality of life and evaluating the performance of cities and their regions, strategic frameworks for regional economic development, social polarization in cities and regions, and issues of urban and regional governance," p. 25. Or: "The objective of [sustainable] development would be human welfare in balance with nature, based on the values of democracy, equality before the law and social justice, for present and future generations, in the absence of ethnic, economic, social, political or gender discrimination or that based on creed" quoted in Carrion (1997). But a much better formulation is found on page 32, which speaks of humanizing the city. To quote Peter Hall: "The late Aaron Wildavsky once wrote a paper with the title 'If planning is everything, maybe it's nothing'. His argument could apply to sustainability as well; it could come to mean anything you think is OK and ought to be done...." in Mega and Petrella (1996). For one of the efforts to broaden the meaning of the term, yet give it a strongly critical meaning, see Bernd Hamm (1992).

[13] Or, to go one step further, listen to the president and chief executive of the empowerment zone, Deborah C. Wright, who said that some of the concerns about the evolving economy of 125th Street are perhaps justified in the eyes of the community. But "... the fact is, " she said, "... capitalism has no plan, except to go where money can be made. ... It's scary, frankly, because, as you know, one of the basic tenets of capitalism is that you can't control it. ... Nor do I think we want to. We want to prepare people to compete in a market based economy because that is the only thing thus far that has been shown to be *sustainable*." Or, "If a neighborhood is to retain stability, it is necessary that properties shall continue to be occupied by the same social and racial classes. A change in social or racial occupancy generally contributes to instability and a decline in values" quoted in USFHA (1938) and McKenzie (1994).

[14] Actually, the term has mixed provenance. On the one hand, it is related to the 'land ethic' of Aldo Leopold, which is frequently cited in treatises on sustainability. See, for example, Lukerman and Nordstrom (1997). On the other hand, it has been expanded frequently into a blanket slogan serving many purposes, as we argue at the end of this paper.

[15] The World Business Council certainly sees 'eco-efficiency' as a profitable, market consistent and indeed market driven, aspect of international business. See De Simone et al. (1998). DeSimone is CEO of 3M and Popoff chairman of

the board of Dow Chemical. Joshua Karliner, (1998) points out, as cited by Neff, that Chevron spent US$ 5,000 on a butterfly protection programme at its El Segundo refinery but spent more than US$ 200,000 producing an ad boasting about it—and el Segundo is one of the largest single sources of pollution in the greater Los Angeles area.

[16] What 'long run' means is, of course, always a matter for debate. In 1992, the United Nations Conference on Environment and Development (UNCED) concluded that time frames should be extended from a few years to a few generations. Cited in Lawrence (1996), p. 46. But any specific definition is necessarily arbitrary.

[17] "... ecological stewardship, social equity, and economic prosperity are the essential ingredients for sustainable human progress" summarizes a review of four leading works on sustainable communities. The statement is more of a postulate than a conclusion. See Lukerman and Nordstrom (1997).

[18] This is an interesting logical question. Is a measure that is not sustainable *ipso facto* undesirable? One argument against the worship of the capitalist system as 'the end of history' is that capitalism is not sustainable in its present form and that there necessarily will be other forms of economic organization replacing it because it cannot continue as it is today. Is that a logical criticism of contemporary capitalism? I think not. It only becomes such if the further argument is made that the negatives of its end will outweigh the positives of its growth. It is, then, not the fact of unsustainability that matters but the consequences that flow from it, a quite different matter. A single person's life is not 'sustainable' indefinitely but that is no reason not to value it.

[19] The point is the same as with the frequent debates about whether a given proposal is 'practical' or not: if practicality becomes a goal rather than a constraint, the result is sheer opportunism.

[20] In the interesting evaluation of projects undertaken by the European Foundation for the Improvement of Living and Working Conditions (1996) the usefulness of such an approach can be seen. Issues such as 'level of crime' are listed as a measure of social sustainability but no distinction is made between long and short-term impacts so that unsustainable measures might well be given a higher rating than sustainable ones, e.g., police crackdowns or long prison sentences *vs* job generation or rehabilitation.

[21] That precise calculation is made when it is decided to finance housing construction through borrowing rather than all at once, up front; more gets built now, even if the certainty of as many being built next year is reduced by the ongoing burden of repayment for past construction. The opposite calculation was made by the Austrian Social Democrats in the 1920s, in deciding to pay for new social housing projects all at once, hoping thereby to make it easier to fund new construction in following years. See Marcuse (1986).

[22] Not only possible, but frequent. The Sustainable Cities Programme on UNCHS/UNEP, for instance, states flatly: "The SCP activities are primarily focused upon promoting more efficient and equitable use of natural resources, and control of environmental hazards in cities..." in Sustainable Cities Program (1998).

[23] I am aware that a conflict between the two principles of low-cost housing and environmental protection can generally be avoided and is often used as a cloak to oppose housing for poor people (see Mary Brooks' work, for instance); nevertheless, the possibility of a conflict is real.

[24] "The Llujiazui International Consultative Process also perpetuated the contradictory approach to 'sustainable development' planning where a designer's concerns rest with reducing energy consumption within a small spatial area while ultimately supporting broader processes, such as the plundering of China's natural resources by financial institutions which use these urban spaces as bases for their 'command and control' activities." (quoted in Olds 1997).

[25] From a review by Kristin Kaul of van der Ryn and Cowan (1995).

[26] The literature, by now, is extensive. See the citations in a recent excellent review, Collin and Collin (1994).

[27] Many have made the same point. For a recent comment in our specific context, see Albrechts (1997).

[28] David Harvey has put forward this argument very eloquently in Harvey (1996); also more recently and concisely, in Harvey (1998) in which he points out that a wing of capitalism is quite content to judge sustainability in terms of the continuity of capital accumulation, and calls for a "... more nuanced view of the interplay between environmental transformations and sociality."

[29] As many definitions do not. See, for instance, the formulation of the Commission of European Communities: "... [sustainable] is intended to reflect a policy and strategy for continued economic and social development without detriment to the environment." Cited in Lawrence (1996) p 65.

[30] See the innumerable calls for us to 'rethink our priorities': "A new ethic must be put into practice. But this will remain impossible unless we stop thinking of our participation in the common good as a tax," states Bertrand Renaud, Head of the Urban Affairs Division, OECD. Or: "The developed countries have to recognize that their urban lifestyles ... are an important part of the global environment problem." Klaus Töpfer, UN Commission on Sustainable Development. Quoted in page iii of Price and Tsouris 1996. The creation of a President's Council on Sustainable Development flows from the political belief that the formulation is a non-controversial, universally accepted one.

[31] A point also eloquently made by Sandra Rodriguez (1998). To quote from this, "An underlying premise in discussions of sustainability is that 'we' are in this together. This generic 'we' assumes that all people are equally to blame for society's environmental problems and that 'we' all have a responsibility to change our lifestyles to 'save the planet'." As Catherine Lerza asks, "Are the poor, the marginalized equally to blame for the waste and pollution that exists, when they are the people least benefiting from economic growth and they are bearing most of the environmental burden?" (idem p. 5).

References

Albrechts L (1997) Genesis of a Western European spatial policy? *Journal of Planning Education and Research,* Vol. 17

Bhatti M, Brooke J and Gison M eds (1994) Housing and the environment: A new agenda. *Housing Studies* Vol. 12, No. 4, Chartered Institute of Housing, Coventry, p. 579

Karliner J (1998) *The Corporate Planet: Ecology and Politics in the Age of Globalization,* Sierra Club, San Francisco CA

Carrion D (1997) Re-thinking housing production: time for responsible co-responsibility. *Building the City with the People,* Habitat International Coalition, San Rafael, Mexico, p. 27

Collin RM and Collin R (1994) Where did all the blue skies go? Sustainability and equity: The new paradigm. *Journal of Environmental Law and Litigation,* Vol. 9, pp. 399–460

De Simone LD, Popoff F, World Business Council for Sustainable Development (1998) *Eco-efficiency: the Business Link to Sustainable Development.* MIT Press, Cambridge MA

European Foundation for the Improvement of Living and Working Conditions (1996) *Towards an Economic Evaluation of Urban Innovative Projects.* The Foundation, Dublin

European Foundation for the Improvement of Living and Working Conditions (1998) *Redefining Concepts, Challenges and Practices of Urban Sustainability,* The Foundation, Dublin

Hamm B and Muttagi PK eds (1992) *Sustainable Development and the Future of Cities.* Centre for European Studies, Trier

Harvey D (1996) *Justice, Nature and the Geography of Difference.* Blackwell, London

Harvey D (1998) Marxism, metaphors and ecological politics. *Monthly Review,* April 1998, pp. 17–31

Lawrence RJ (1996) Urban environment, health and the economy: cues for conceptual clarification and more effective policy implementation. In: Price C and Tsouros A, eds. *Our Cities, Our Future: Polices and Action Plans for Health and Sustainable Development.* WHO Healthy Cities Project Office, Copenhagen

Lukerman BL and Nordstrom R (1997) Sustainable communities. *Journal of the American Planning Association,* Vol. 63, Autumn, p. 513

Marcuse P (1974) Conservation for whom? In: Smith JN (ed) *Environmental Quality and Social Justice in Urban America.* The Conservation Foundation, Washington DC, pp. 17–36; reprinted in *California Today* Vol. 2, No. 6, June 1974

McGranahan G, Songsore J and Kjellen M (1996) Sustainability, poverty and urban environmental transitions. In: Pugh C, ed. *Sustainability, the Environment and Urbanization.* Earthscan, London, p. 103

McKenzie E (1994) *Privatopia: Homeowners Associations and the Rise of Residential Private Government.* Yale University Press, New Haven CT, p. 57

Mega V (1996) Fragments of an urban discourse. In: *Utopias and Realities of Urban Sustainable Development,* Conference Proceedings, Turin, Barolo, September 1996, pp. 66–67

Mega V and Pedersen J (1997) *Urban Sustainability Indicators.* European Foundation for the Improvement of Living and Working Conditions, Office for Official Publications of the European Communities, Luxembourg, p. 2

Neff G (1997) Greenwash. *The Nation,* November 1997, p. 50

Olds K (1997) Globalizing Shanghai: the 'global intelligence corps' and the Building of Pudong. *Cities,* Vol. 14, No. 2, pp. 109–123

Price C and Tsouros A eds (1996) *Our Cities, Our Future: Policies and Action Plans for Health and Sustainable Development.* World Health Organization, Copenhagen

Rees W (1988) A role for environmental impact assessment in achieving sustainable development. *Environmental Impact Assessment Review,* Vol. 8, p. 279

Rodriguez S (1998) Sustainable and Environmentally Just Societies. *Planners' Network,* No. 129, May 1998, pp. 4–7

Satterthwaite D (1997) Sustainable cities or cities that contribute to sustainable development? *Urban Studies,* Vol. 34, No. 10, pp. 1667–1691

Sustainable Cities Program (1998) *Sustainable City News,* Vol. 1, No. 4, June 1998, p. 2

van der Ryn S and Cowan S (1995) *Ecological Design.* Island Press, Washington DC

WCED: World Commission on Environment and Development (1987) *Our Common Future,* The Brundtland Report. Oxford University Press, New York

Wildavsky A (1996) If planning is everything, maybe it's nothing. In: *Utopias and realities of urban sustainable development,* conference proceedings, Turin, Barolo, September 1996

USFHA: United States Federal Housing Administration (1938) *Underwriting Manual: Underwriting and Valuation Procedure Under Title II of the National Housing Act.* US Government Printing Office, Section 937, Washington DC

Sustainable Development and Urbanization

MARIOS CAMHIS[1]

Introduction

During the last 30 years we have been witnessing an increasing awareness of the interrelationships between the economic, social and environmental dimensions of development. Three key words could describe the new paradigm that has emerged:

- **Acceleration:** We have been experiencing changes at a pace that was unimaginable in other periods in history: the information society, the population explosion, rapid urbanization, and climate change are the most characteristic examples.

- **Human impact**: Up to now, nature's absorbing or carrying capacity could withstand the impact of human activities. Today they affect global cycles and systems.

- **Internationalization**: The economy is global. Environmental problems do not recognize borders. What happens in one place has adverse effects somewhere else, ranging from trans-boundary to global impacts. The sum of individual actions can have different global results, in qualitative terms, from its components.

At the same time, considerable efforts have been made to assist developing countries in achieving economic growth. The Bretton Woods Institutions (World Bank, IMF, WTO), specialized bodies of the UN and the major country donors (USA, EU, Japan) developed policies and disbursed funds and technical assistance, which did not always have the intended results. Despite global economic growth, the gap between rich and poor has widened. Approximately 850 million people in industrialized countries have experienced dramatic improvements in their environments. These countries are also the main beneficiaries of global economic development. Increased wealth translates into a better local environment but a larger

69

M. Keiner (ed.), The Future of Sustainability, 69–98.
© 2006 *Springer. Printed in the Netherlands.*

contribution to greenhouse gas emissions. The 'ecological footprint'[2] of industrialized countries has by far exceeded their relative population size and land area.

"The richest 20% of the world's population accounts for 86% of total private consumption expenditure, consumes 58% of the world's energy, 45% of all meat and fish, 84% of all paper, and owns 87% of cars. Conversely, the poorest 20% of the world population consumes 5% or less of each of these goods and services." (UNDP 2001)

The remaining 5 billion of the world population face serious problems: environmental degradation, bad practices and corruption, increased and uncontrolled urbanization, industrialization with old polluting technologies, and poverty and poor health. The figures are startling: 1 million/year die from urban air pollution; 2 million/year die from exposure to stove smoke in houses; 3 million/year die from water-related diseases; 1.2 billion live on less than $1 per day; 3 billion live on less than $2 per day; 1.5 billion lack access to safe water; 800 million (200 million children) are suffering from chronic malnutrition; 68 million will die from AIDS by 2020 (55 million in sub-Saharan Africa alone).

In the context of these processes, the role of cities and urban areas is significant but underestimated. Cities have been the motors behind economic development in industrialized countries. Urbanization and GDP growth have been synonymous. Cities have undergone dramatic improvements in their environment, but they are also the main sources of greenhouse emissions. In developing countries, the cities are the most striking expression of poverty concentration, crime, and environmental deterioration. These problems will be multiplied in the next decades, if appropriate action is not taken at all economic, environmental and social levels. The challenges and the rates of change are of such magnitude that past 'Western' models might not bring the expected results. Urbanization in the developing countries will not necessarily lead to GDP growth and the old recipe of 'grow and pollute now, pay and clean later' might not have the time to mature and pay dividends.

The promotion of sustainable urbanization is a key to global sustainable development. A more effective and environmentally friendly development policy should be more spatially oriented, giving particular emphasis to the problems of cities. It should also be accompanied by the innovative use of new technologies and the strengthening of governance structures. International actors will have to redirect their attention to the urban problems of the developing world and significantly increase their coordination efforts.

The Challenges and the Problems

The world population increased from 2.5 billion in 1950 to 3 billion in 1960 and to 6 billion in 2000. It will further increase by 8 (low estimate) to 13 (high estimate) billion by the year 2050. The geographical distribution of this increase is unequal. The totality of the increase will occur in developing countries, with the notable exception of the USA. India will grow from 1,050 to 1,628 million and China from 1,282 to 1,394 million. Pakistan (from 144 to 332 million) and Nigeria (from 130 to 304 million) will more than double and the DR of Congo and Ethiopia will triple their populations (UN Population Division 2001a).

Linked to population changes are three important phenomena: migration, ageing, and urbanization. The number of immigrants worldwide has more than doubled since 1975. Around 3% of the world's population (175 million) resides outside the country of their birth: 60% in the more developed regions and 40% in less developed ones (UN 2002a). Immigrants tend to concentrate in the major cities. Migration has important economic consequences both for the countries of origin and the receiving countries. Workers' remittances have been an important addition to the GDP of a number of countries often by more than 10%. On the negative side, migration leads to the drainage of the most dynamic elements from the countries of origin, benefiting the most developed areas (UN 2002a). In the developed countries the ageing process has taken on threatening dimensions resulting in significant difficulties for many countries to maintain their welfare system. This is a well-known fact. What is less evident is that ageing will also become a problem in the developing world due to a sharp drop in fertility levels and rapid increase in life expectancy. By 2050, the number of elderly people in less developed countries is projected to more than quadruple (from 374 million in 2000 to 1,570 million). Asia and Latin America are ageing most rapidly, and the elderly will make up 20 to 25% of their populations by 2050 (European Commission 2002a). The impact of the ageing process will be an additional burden to cities in developing countries.

The major challenge of the 21st Century is the dramatic change in the spatial distribution of the population. The world will be characterized by unprecedented rates of *urbanization*. In 2000, the world's urban population was 2.9 billion; it will increase to 5 billion by 2030, that is, 60% of the world's population. At current rates of change, the number of urban dwellers will equal the number of rural dwellers in the world already in 2007 (UN Population Division 2001a). Virtually all the population growth of 2.2 billion expected at the global level during the period 2000–2030 will be concentrated in the urban areas of the less developed regions. Their urban population is likely to rise from approximately 2 billion in 2000 to just under 4 billion in 2030 or by 60 million people a year—equivalent to the population of Egypt or Ethiopia (UN Habitat 2002, World Bank 2003).

Urbanization levels in developed and developing regions are gradually converging. In the less developed regions, 40% of the population lived in urban areas in 2000 compared with only 18% in 1950. This number will reach 56% by 2030. Latin America and the Caribbean are already highly urbanized at levels similar to developed regions (75% in 2000, 84% in 2030). In Africa and in Asia, 37% and 48% of their populations respectively, was living in urban areas in 2000. These percentages will rise to 53% and 54% by 2030 (UN Population Division 2001b). Percentages are often misleading. Despite their higher levels of urbanization, the combined number of urban dwellers in Europe, Latin America and the Caribbean, Northern America and Oceania (1.2 billion) is smaller than the number in Asia today (1.4 billion), one of the least urbanized major regions of the world. By 2030, Asia and Africa will each have higher numbers of urban dwellers than any other major region of the world (ibid).

Between 2000 and 2015, some 200 million additional people will have to be accommodated in African cities (ibid), and Asian cities will receive an additional 590 million (UN-Habitat 2002b). Millions of people will be added to those already living in unacceptable situations in many of the world's urban areas. In already urbanized developing countries undergoing severe economic crises, such as in Latin America, previously prosperous neighborhoods are being transformed into slums. Two examples suffice to illustrate this point. Nouakchott in Mauritania has grown in 30 years from 40 to more than 600 thousand people. More than 40% of urban land is occupied by squatter settlements and the proportion of individuals living below the poverty line is estimated at 50% (Cities Alliance 2000). In São Paulo, home to 18 million people, the downtown core has lost many businesses and residents to newer business districts and the outlying suburbs. The favelas continually expand as waves of poor people from elsewhere in Brazil arrive to build makeshift homes on undeveloped land at the city's edge (Zwingle 2002). A recent study has shown that around 1 million of new households are being added each year in Brazil, most of them seeking shelter in the favelas (World Bank 2002).

Very large urban agglomerations tend to attract more attention, but the proportion of world population living in mega-cities of 10 million inhabitants or more was only 3.7% in 2000 and will increase to 4.7% by 2015. What is happening is that the number of such cities will keep rising. By 2015, Tokyo will remain the largest urban agglomeration with 27.2 million inhabitants, but it will be followed only by cities in developing countries: Dhaka, Mumbai (Bombay), São Paulo, Delhi and Mexico City, all of which are expected to have more than 20 million inhabitants. Just 9 out of the world's 40 cities with 5 million inhabitants or more in 2001 were located in developed countries, and the equivalent figure will be 10 out of 58 in 2015 (UN Population Division 2001a). Overall, the largest shares of the world urban population increase will be attributed to urban settlements with fewer than 500,000 inhabitants (44.4%) and cities with a population ranging between 1 and 5 million inhabitants (ibid). Today, almost 400 cities have one million people or more. Three quarters of them are found in low and middle-income countries.

In the developing world, small and medium-size cities risk to experience the same problems and pressures as larger urban areas. In some countries, urban air quality tends to be worst in medium-size cities with populations between 100,000 and 500,000 people (Lvovsky 2001). Very often, small urban areas of 100,000 and below lack adequate provision in terms of drinking water and electricity supply, waste disposal, and schools (Montgomery et al. 2003). Crime and vulnerability to disasters often become problems in urban agglomerations well below 1 million inhabitants, but do not increase proportionally with population size. Congestion tends to worsen with city size but is also influenced by other factors such as public transport, traffic management, and road space. Multimillion inhabitant cities are not necessarily the best nor the worst cases of sustainable development. Many of the economic benefits of urban productivity, such as higher wages and increased human capital, appear positively correlated with city population, at least to a fairly high threshold (World Bank 2003).

Rapid urbanization will have significant impacts on the economies, societies, the natural resources, and the environment of the countries concerned. Urban areas tend to suffer more from *fresh water* scarcity. Governments and international agencies have underestimated the number of urban dwellers who have inadequate provision for water and sanitation and the very serious health consequences that this brings for hundreds of millions of people (UN-Habitat 2003). Wasting scarce fresh water resources through bad management is more likely to take place in cities than in rural areas. Mexico City has an abundance of water resources but faces serious problems of water waste and overdraft, which can be resolved with modest policies of better management (Connolly 2001). Contrary to the dangers of water shortages, *food and agriculture* have undergone major changes in recent years. Demands on food have been growing commensurate with population growth. The problem has not been one of quantity but its unequal distribution. Achieving food security requires an abundance of food, access to that food, nutritional adequacy and food safety (FAO 2001). For burgeoning urban populations, the situation might be more difficult in the future. Olivio Argenti, a FAO expert in urban food security, argues

"... urbanization is likely to eat up the productive land, pushing food production further and further away. This increases the cost of all activities associated with producing food and bringing it to cities, calling for massive investments. The consequences are all the more critical where infrastructure and services such as transport, storage, slaughterhouses and markets are already overstretched, which is the situation in most cities in developing countries." (Argenti 2002)

Microbial food contaminants are common, especially in urban areas where food must travel long distances before consumption. The poorest are the most likely victims (FAO 2001). Developments in *drugs* and protective measures to combat AIDS and malaria have been significant. Pricing and distribution prevent their effective use by the less favored segment of world population.

Climate change will continue to be a key international policy issue in the coming years. The US and the EU, the most important contributors of greenhouse gas emissions, either do not want or cannot change their production or consumption patterns at least in the short term. It is a well-known fact that the US has refused to ratify the Kyoto Protocol. Instead, the US government supports a number of Climate Change-Related Programs for hydrogen fuel cell research. An analysis by the Energy Information Administration concluded that even if carbon intensity is cut by 1.5% a year, carbon emissions will still grow about 1.5% a year because of expected economic growth (White House 2002). The EU has ratified the Kyoto protocol. This has not been enough. The EU has failed to reduce CO_2 emissions (European Environment Agency 2003). The overall picture of the situation in 2030 is pessimistic. In relation to 1990 figures, the US's contribution to CO_2 emissions will increase by 50%, compared to an 18% EU increase. In addition, "developing countries are expected to have a serious influence on the global energy picture, representing more than 50% of the world's energy demand, as well as a corresponding level of CO_2 emissions." (European Commission 2003) The countries in the process of development have few incentives and no effective support to take the appropriate measures. Chinese CO_2 emissions have grown explosively since 1950 (Sundt 1999). The centers of economic activity are the cities. In the developing countries, we will witness the cumulative impact of the deterioration of their ambient air quality and the increase of greenhouse gas emissions (Leitmann 2003). Cities contribute to climate change but they could also be the prospective victims of this phenomenon. Little research has been carried out on the impact of climate change on cities. Urban areas with inadequate infrastructures will run higher risks from increased storms, flooding, mudslides and soil erosion. Coastal mega-cities of the developing world will be most threatened from rising sea levels (World Bank 2003). The future of the Earth's environment will be decided in the cities and, in particular, in Asia with 60% of the population and the fastest growing economies (Fahn 2003).

Cities suffer the most from any negative impacts of *globalization*. The benefits and costs of economic growth have not been spread evenly. In many countries, real incomes have fallen, living costs have gone up and the number of poor households has grown, especially in cities (UNCHS 2001a). The process of global economic restructuring has often led to "the urban and social fabric of most cities becoming more and more fragmented and stratified." (Recife Declaration 1996)

"The social and economic cores and peripheries of the global information age and the global economy are not only continents apart but can now be found geographically adjacent to each other in individual cities."[3]

In many cases, living conditions are not simply worsening but becoming unlivable. Historically, there has been a strong, positive link between national urbanization and national levels of *human development*. Urban population, as a share of total national population in both highly industrialized countries (HIC) and those countries with a high Human Development Index (HDI, is above 70%. Urbanization falls to less than 30% in countries that are classified as Least Developed Countries (LDC) or have a low HDI (UNCHS 2001b). Development and urbanization appear to proceed hand-in-glove. Present trends indicate that this relationship might change in the future. Africa is a case in point. Africa's experience over the last 30 years has been one of urbanization without growth. This is a unique phenomenon, even across poor countries and poor growth performers. From 1970 to 1995, the average African country's urban population grew by 5.2% per annum while its GDP declined by 0.66% per year (Hicks 1998).

High urbanization rates in the developing countries are accompanied by a dramatic increase in the number of poor.

"Urban poverty is growing in scale and extent, especially at the peri-urban rim. In Latin America, Europe and Central Asia, more than half the poor already live in urban areas. By 2025, two-thirds of the poor in these regions, and ... almost half the poor in Africa and Asia will reside cities or towns." (World Bank 2000b)

The number of poor and malnourished children in urban areas is growing in a number of countries, including India and China with increased rates of growth (IFPRI 2000). Urban poverty increased in all East Asian crisis countries. In Korea, for example, poverty among its urban population more than doubled from 9% in 1997 to 19% in 1998.[4] In this context a major issue is the particular nature of urban poverty. Still today, poverty in rural and urban areas is measured on the same basis. The measurement for both cases is per capita income of less than 1 or 2 dollars per day. But poverty in rural and urban areas is not the same. There are several interrelated factors contributing to urban poverty: lack of employment, income, and social security; lack of access to credits for business or house; inability to afford adequate housing; tenure insecurity; unhygienic living conditions; and poor health and education. They all contribute to a strong sense of insecurity and disempowerment.[5] A number of studies are beginning to explore this issue, showing that higher income in urban areas does not necessarily result in a higher standard of living.

Although poor city dwellers usually live closer to *health facilities*, safe water supplies, schools, and sanitation facilities compared to rural people, they often cannot afford to use these services. Unsanitary and overcrowded conditions contribute to increased illness and mortality among children and diseases among adults, threatening their ability to work and support their families (IFPRI 2000). Some classify everyone who has a *water* source within 200 meters of their home as having an adequate supply of water. Having a public tap within 200 meters for 200 people in a rural settlement is not the same as having a public tap at the same distance for 5,000 people in an urban area (UN-Habitat 2003). Urban dwellers tend to purchase most of their *food*, while rural people grow at least some of their food. In urban areas, people spend an average of 30% more on food than in rural areas but they consume fewer calories.[6] The majority, if not all, of poor urban dwellers are in an insecure *housing* situation, living in rented or self-built illegal dwellings. Their home is often the base for household enterprises and a foundation for an entire network of social support. Eviction threatens the mechanisms by which the poor survive in cities. Urban dwellers spend a significant amount of their income on *transport*. This percentage is even higher for the urban poor who, for the most part, live at the periphery of cities, where they find land to build their illegal housing.

Extreme poverty in urban 'proximity' is potentially much more dangerous than rural poverty. "The poor come to understand exactly what it is they are lacking." (Stephens 2002) Child criminality is much higher in cities. Both parents are absent long hours, which make it difficult for them to care for their children. The potential for *social violence* and urban unrest seems greater than ever before. A form of urban war is also emerging. Comuna 13, a poor, sprawling neighborhood at the western edge of Medellin has been the scene of the Colombian military's largest urban offensive against guerrilla supporters, with 3000 troops backed by helicopters (Wilson 2002).

The condition of the urban poor is worsened by their spatial concentration in *slums* under various names, such as *favelas, kampungs, bidonvilles,* and *tugurios.*

"[They range] from high density, squalid central city tenements to spontaneous squatter settlements without legal recognition or rights, sprawling at the edge of cities. Some are more than fifty years old; some are land invasions just underway. Slums lack basic municipal services, water, sanitation, waste collection, storm drainage, street lighting, ... [and] roads for emergency access. They are usually far from schools and clinics and do not have safe areas for children to play ... While the average age of city populations is increasing, the average age of slum dwellers is decreasing, so youths and children suffer most." (World Bank 2001)

Estimates by UN-HABITAT show that as many as 712 million people lived in urban slums in 1993. In 2001, based on data from 232 cities, the number of people living in slums was projected to be 837 million. UN-HABITAT estimates that 56% of Africa's urban population is now living in slum conditions.[7] These figures are both underestimations and rapidly growing. Recent estimates raise this figure to 900 million. It could be 1.5 billion by 2020 (UN Millennium Project 2004). In developing countries, slums comprise nearly all of the urban areas, which accommodate incoming populations. Urbanized countries in Latin America see an important deterioration in the conditions of their cities and the transformation of whole areas into slums.

"In terms of supply, the number of dwellings constructed annually in the Third World is usually between 2 and 4% per 1,000 inhabitants while the population is expanding at between 20 and 35 and the urban population at between 25 and 60 persons per 1,000 inhabitants a year." (World Bank 2002)

In India, a demographer summarized the situation as the '2-3-4-5 Syndrome'. In the last decade, India grew by 2%, urban India by 3%, mega-cities by 4% and slum populations by 5% (Chatterjee 2002). The comparison of these facts with the relevant Millennium target of improving by 2020 the lives of 100 million slum dwellers is consternating. "The problem of current slum and squatter settlements is only a glimpse of the future." (MIT 2001)

During the next decades the situation will worsen. A number of factors point in this direction: increase in population in developing countries and concentration of this increase in urban areas; slow economic growth and/or incapacity to meet the needs of the existing and future poor population;

development policies, which have not been adapted to cater to the urban poor.

The Policies of International Actors

A considerable number of International Actors are involved directly or indirectly in defining environmental and development policies. They often have divergent objectives and approaches. Multinational meetings, in particular in the framework of the UN, have contributed to the identification of common objectives and convergence of ideas. During the last decade we have also witnessed a significant, though slow, change in the policies of international actors both as regards the integration of environmental concerns into the development process and the increased efficiency of development policies. Today, actors taking part in global governance have a clearer set of objectives than in the past. They operate under a revised set of policy instruments but with little coordination and insufficient resources to achieve these objectives.

Setting *objectives* is important to provide a framework for action. It does not necessarily imply that these objectives will be successfully implemented. In September 2000 at the Millennium Summit, world leaders agreed on the 'Millennium Development Goals', most of which have the year 2015 as a timeframe and use 1990 as a benchmark (UN 2000). These goals are both modest and ambitious. They are thematic and have little spatial differentiation. They include halving the proportion of people living on less than a dollar a day and those suffering from hunger; achieving universal primary education and promoting gender equality; reducing child mortality and improving maternal health; reversing the spread of HIV/AIDS; integrating the principles of sustainable development into country policies; reducing by half the proportion of people without access to safe drinking water. The urban dimension is confined to achieving significant improvement in the lives of at least 100 million slum dwellers (by 2020). In September 2002 the Johannesburg Summit reconfirmed the Millennium goals and complemented them by setting a number of additional ones such as halving the proportion of people lacking access to basic sanitation; minimizing harmful effects from chemicals; and halting the loss of biodiversity (UN 2002b).

It is generally recognized that the *resources* available to achieve these objectives are insufficient. Current aid flows are only around $50 billion per year, ($60 billion in 1990), half of which is being provided by the European Union. Private sector investments have increased during the last decade but the levels are fluctuating: they amounted to $30 billion in 1990, $300 billion in 1997 and $150 billion in 2002 (Wolfensohn 2003). Future prospects are uncertain. In 2000 in the *Monterrey Consensus*, various Heads of State and Governments stated that it is imperative to meet the challenges of financing development and that there is a need for a global response. They admitted that there are "dramatic shortfalls in resources required to achieve the internationally agreed upon development goals (including the UN Millennium Declaration)" and pledged, among other things, to mobilize and increase the effective use of financial resources; mobilize domestic resources; promote international trade; support policies implemented with the full participation of developing countries; and promote a holistic approach to the challenges of financing (UN 2002c). Up to now, the response to the Monterrey Consensus has been modest.

During the 1990s, a certain convergence of ideas among donors and international organizations was reached regarding the most effective *implementation* approaches. Important aspects of this effort include the local ownership of strategies and reforms in the institutions of governance, moving to a more participatory process. As a result, many aid agencies have introduced the concept of results-based management (RBM). Capacity building has also been a new key element of development policies. Capacity means the ability of individuals, organizations and societies to perform functions, solve problems, set and achieve goals. Capacity development entails the sustainable creation, utilization and retention of that capacity, in order to enhance self-reliance and reduce poverty (OECD 2001).

The World Bank introduced the Comprehensive Development Framework: an approach by which countries can achieve the objectives of eliminating poverty, reducing inequalities by emphasizing the interdependence of all elements of development—social, structural, human, governance, environmental, economic and financial. It advocates a holistic long-term strategy in which a country owns and directs the development agenda with the Bank and other partners, each defining their support. Operationally this is translated into the Poverty Reduction Strategy Paper (PRSP), an annually updated strategy document that each country prepares in collaboration with the World Bank and the IMF (World Bank 1999).

In the *US*, the Millennium Challenge Account (MCA) of 2003 pledged to promote a

"... new partnership between all parties involved in success development: donor and recipient governments, non-governmental and private voluntary organizations, business and multilateral organizations...." *(USAID 2003)*

The US intends to channel additional funds only to developing countries that demonstrate a strong commitment to ruling justly, rooting out corruption and protecting human rights and political freedoms. They should also invest in their people and encourage economic freedom. Investments will be targeted to agricultural development; enterprise and private sector development; governance; health; trade; education and capacity building (ibid).

At the Barcelona European Council of March 2002, the Heads of States of the *European Union* committed themselves to increase their development assistance from 0.33% to an average of 0.39% of their Gross National Product (GNP) by 2006, as a step towards the reaffirmed 0.7% target. This would mean that an extra $20 billion over the period 2000 to 2006. Already the EU accounts for more than 50% of all official development assistance worldwide totaling $25.4 billion in 2000. The main objective of the EU's Development Policy is to reduce poverty. Key to its success is the ownership of the strategies by the partner countries and the most wide-ranging participation of all segments of society; the promotion of a mode of development centered on social and human aspects and on sustainable management of natural resources; and capacity-building and good governance for transparent management of all resources.[8] In order to maximize the impact of Community development policy, Community activities will be refocused in a limited number of areas, namely: link between trade and development; support for regional integration and cooperation; support for macro-economic policies; transport; food security and sustainable rural development; and institutional capacity-building. Better synergy will also be sought both within the Union and with other donors (ibid).

Five years after setting the Millennium Development Goals, progress is still very slow. If no immediate action is taken we are going to fall far short of the announced targets for addressing extreme poverty. This is the main conclusion of a UN report made public in January 2005. The report presents the findings and recommendations of the UN Millennium Project an independent advisory body to UN Secretary-General (Sachs 2005). Official Development Assistance has been up to now inadequate. It has at least to be doubled reaching $135 billion in 2006 and $195 billion in 2015 that is 0.44 and 0.54 percent of donor GNP, still less than the 0.7 percent promised in Monterrey. This is one of the ten key recommendations of the report which also include among other: opening of high-income countries markets through the Doha trade round; strengthening of the coordination of UN agencies, International financial institutions, funds and programs; and a group of Quick Win actions to improve millions of lives such as the free distribution of malaria bed-nets and medicines for all children in affected regions (op cit pp. xiv–xvi).

Strengthening the Urban
Dimension of Development Policies

Population growth, the persistence of inequalities, poverty, health prob-lems, and environmental disasters are being tackled with relatively modest development policies and financing. The new challenge of rapid urbaniza-tion in the developing world has not yet drawn the appropriate attention of international actors. In an urbanizing world where large cities are develop-ing very quickly, urban poverty and the management of metropolitan areas are among the major challenges of this century. If no action is taken at dif-ferent levels, the negative effects can be enormous. There can be no sus-tainable national economic development without economically strong and properly functioning cities. Aid organizations tend to ignore the fact that as an engine of social and economic development, the city can also contribute to sustainable rural development. Despite general declarations agreed in the different specialized international meetings of the UN and some recent activities of the World Bank, the urban dimension of development policy is still very weak.

It has been argued that in Africa, aid agencies have an anti-urban bias.

"Several international development agencies in Africa still have no de-partment specifically in charge of urban development. In several agencies, the ruralist lobby is so strong that urban poverty is hardly recognized as such." Urban development is disguised behind "the imperatives of health, education, gender, family planning, micro-enterprise promotion, environ-ment...." (UNCHS 2001b)

This problem is not only confined to Africa.

"Sustainable urbanization remains marginal in terms of both the re-sources involved and mainstreaming within development assistance policy ... attributable partly to the nature of international assistance, which tends to focus on sector-specific issues ... and partly to the long-lasting and pre-vailing anti-urban bias on the part of donor and international agencies alike." (UN-Habitat 2002b)

"The main international urban cooperation programs, such as in trans-port, sanitation, and water supply have been fragmented and often politi-cally, socially, and technologically unsustainable, even in the short-term." (Atkinson 2002)

This reticence is strengthened by the fact that reducing poverty and im-proving living conditions can be more difficult in urban areas than in rural ones.

"Policymakers and aid officials frequently know what tools and programs they can use to promote social and economic development in rural areas, where agriculture is a key. But the urban environment is more complex and diverse." (Garrett and Ruel 1999)

One of the main reasons given for the lack of urban focused programs is that poverty remains higher in rural than in urban areas. A recent document of the European Commission states,

"With poverty reduction as the central objective of EC development policy, there is a need to address more systematically and in a more comprehensive manner rural development concerns, because poverty and hunger are mainly rural problems." (European Commission 2002b)

But as we have just seen, urban and rural poverty are different. Even based on the same measurement methods, the number of urban poor will surpass the number of rural poor in the next 15–30 years. There is also an important discrepancy between global figures given by International Organizations from which we deduce a much lower percentage of urban poor than what emerges from specific city studies, as in the example of Brazil. Today 82% of the Brazilian population lives in urban centers. The urban poor, in turn, account for 74% of the total Brazilian poor (World Bank 2002).

In the run up to the Johannesburg Summit in September 2002, there was a sudden proliferation of publications on sustainable urbanization, supported mainly by UN agencies. The 'State of the World's Cities', for example, emphasized the need for national governments to develop National Urban Policies (UNCHS 2001b). Such policies should, among other things, institute participatory national planning and budgeting processes reflecting local level priorities; integrate physical with economic planning; recognize urban regions as geographic planning modules; devolve service provision and revenue raising; and develop local capacities to take on new functions such as in urban planning. Local Agenda 21 and the Habitat Agenda each identified a number of very similar guiding principles for urban sustainability including strategies for sustainable energy and transport; adequate shelter for all; conservation of historical and cultural heritage; combating poverty and community empowerment; and promoting local labor, intensive economic growth, and responsible fiscal policies (UN-Habitat 2002a).

UN agencies have compiled an important number of case studies covering mainly the fields of environment, housing and government.[9] Such best practices often have limited impact either because they are too small, compared to the scale of the problems and/or they have been conceived as a one-shot operation with no adequate follow up and monitoring. The constraints impeding the successful implementation of urban projects include the lack of adequate resources; a non-spatial and/or integrated approach; a resistance of national administrations to support decentralization; the inadequate use of new technologies; the failure of local authorities to keep in pace with rapid urbanization and the magnitude of related problems. Darshini Mahadevia, in reviewing urban development initiatives in India, argues that all efforts undertaken by central or local governments were rarely conceived with a view to the possibilities of mutual reinforcement or synergetic interaction. W. J. Kombe reviews the Sustainable Dar es Salaam Project launched in 1992 with the support of UN-Habitat (UNHCS) and UNEP. The project established working groups envisaging a wide participation of stakeholders to propose solutions to the most pressing environmental problems. While working groups appear to have mobilized new collective forms of problem solving, their most important proposals could not be implemented.

"Vested interests among stakeholders, institutional inertia, bureaucratic in-fighting, and a lack of political will at the central level all stood in the way." (Westendorff and Eade 2002)

On the positive side, the relevant literature has repeatedly referred to the good example of sustainable urban management in the city of Curitiba in Brazil. Curitiba grew from a town of 300,000 to a metropolis of 2.3 million. The city managed to overcome its problems by giving particular emphasis to integrating transportation and urban planning. Key to the success of Curitiba has been "a responsive, democratic government oriented toward public participation," with incentives for the urban poor "to participate in housing, street cleaning and recycling programs" and measures for "publicly subsidized loan programs for low-cost housing, licensing of street vendor activity and work opportunities for street children." (Harris 2001)

A number of conclusions can be drawn from the relevant literature. At the theoretical/normative level there is an overall agreement on what the best policies, the types of infrastructures needed, and the appropriate environmental, training and social measures are. A number of best practices have been implemented during the last decade, but most of them have focused on sectoral approaches. There are a number of often insurmountable constraints for the implementation of the 'appropriate policies'. Many international financial organizations or bilateral donors have promoted inappropriate policies and have acted in a non-coordinated way. The overall international climate as regards international macroeconomic policy and international competition is not always conducive to sustainable urban development.

Among international agencies, the World Bank has taken a number of steps to more actively promote an integrated approach to urban problems.[10] An interesting example is the Caracas Slum-Upgrading Project, which aims to improve the quality of life for the inhabitants of a selected number of barrios by developing and implementing a community-driven, sustainable, and replicable infrastructure improvement program. The project has three components. The first is *urban upgrading*. It finances the design and implementation of Neighborhood Improvement Plans (NIP) and includes designing and constructing pedestrian and vehicular access, water distribution, sewerage and sanitation, drainage, electricity distribution, public lighting, and community centers as well as building new houses for resettlement. The second component, *institutional development*, finances the start-up and operational costs for the project management unit, including public dissemination, monitoring and evaluation, and technical assistance and capacity building in several areas. The third component finances the development and operation of a market-based *housing improvement loan fund* which will provide consumer credit to low-income individuals residing in the barrios to finance improvements to their housing unit through a partnership between banks and non-governmental organizations.[11] In 2003, a $100 million loan for the Bogotá Urban Services Project was granted, aimed at improving urban livability by enhancing access, coverage, quality, reliability and inter-agency coordination in the provision of transport, water, sanitation and related basic services, particularly for residents in low-income areas. Among the beneficiaries are around 600,000 low-income residents who live in 14 of the city's poorest zones *(Unidades de Planificación Zonal)*. At a more general level, the Africa Region of the World Bank has identified urban productivity as a key issue for sustainable poverty reduction in Africa.

"A better understanding of Africa's urban economies and urbanization process is clearly needed, in order to identify the role that urban centers need to play in the growth process, and the policy instruments best suited to encourage it, and possible remaining distortions that promote urbanization without growth." (Hicks 1998)

The *Cities Alliance* was launched in 1999 with initial support from the World Bank and the United Nations Centre for Human Settlements (UN-Habitat). Ten governments (Canada, France, Germany, Italy, Japan, the Netherlands, Norway, Sweden, the UK and the US) and four leading global associations of local authorities are also participating. The Asian Development Bank joined the Cities Alliance in March 2002 (Cities Alliance 2002). The European Commission is not participating. Cities Alliance is a global alliance of cities and their development partners committed to improve the living conditions of the urban poor through action in two key areas:

- **City Development Strategies** (CDS)[12] which link the process by which local stakeholders define their vision for their city, analyze its economic prospects, and establish clear priorities for actions and investments;

- **Citywide and nationwide slum upgrading** to improve the living conditions of at least 100 million slum dwellers by 2020 in accordance with the Cities Without Slums action plan (stemming from the Millennium goals) (Cities Alliance 2002).

The Alliance acts as a catalyst, providing only seed funding to help partners to develop slum upgrading programs and/or CDS. By now, well over 100 CDS must have been undertaken, but there were many different definitions. Some have been major analytical plans; others are not much more than mayoral wish lists. Some have been major exercises in consulting the citizens, others more perfunctory (Harris 2002). What the experience of Cities Alliance has in effect revealed is that there is little, if no, coordination on the ground, neither between the different donors nor between the donors and the national and/or local authorities concerned. Each donor international organization or country has to monitor the implementation of its own individual project, relying on the local, regional or national authority concerned for the coordination needed. This is a very serious obstacle to the effective implementation of development programs.

USAID supports both sectoral projects with an urban impact as well as development programs for the urban poor. It was only during the last three years that USAID has been taking a broader approach to urban issues, simultaneously covering housing, education, economic growth and governance. Sectoral projects with an urban impact account for 25% of the total aid, but only about 11% of USAID funds go to the urban poor. This is a very low percentage considering the proportion of poor living in urban areas. For example in the countries of Latin America where USAID intervenes, urban poverty is of the order of 45%.[13]

Urban deprivation has not been part of the agenda of the *European Commission's* 'Fighting Poverty' program. The issues addressed include governance, peace and social stability; rural development and food security; trade; social services delivery; protecting the environment; and sustaining vital transport systems (European Commission 2001). Sectoral projects might have a direct or indirect impact on cities, but they were not conceived for urban areas as such. An explicit urban dimension has been practically confined to two programs of experience exchange between city authorities. The URB-AL, launched in 1995, and 'Asia Urbs Program', established in 1998,[14]

aim to develop links between European and Latin American and Asian local communities respectively through the dissemination and application of
best practices in urban policy fields. These exchanges of experiences have
not been transformed into mainstream policies. A number of Member
States engaged in development policy, such as UK and the Netherlands,
tend to take a stronger interest in supporting urban programs. It is a positive sign that, for the first time, the European Commission in its recent
Communication on the "EU Strategy for Africa" makes an explicit reference to such an approach. It is stated, among other, that, "Africa's demographic boom, rapid urbanization and large scale migration pose new
challenges...therefore, a more integrated approach to sustainable urban
development is needed, based on the twin pillars of good urban governance and good urban management, plus better territorial development and
land use planning..."

The Way Ahead

Making a city work well is very often seen as a luxury of rich developed
countries. Effective urban planning is seen as a consequence of economic
growth rather than as one of its underlying factors, whereas an appropriate
urban strategy can contribute positively to economic development. Development macro-economists emphasize the need for cooperation between
urban planners and macroeconomists today. At an Urban Research Symposium organized by the World Bank in December 2002, Jeffrey Sachs argued that it is important 'to make the urbanization process work' and
'make urban areas true engines of growth'. The way ahead is an integrated
approach, bringing together urban planning, an urban development strategy and efficient urban governance (Sachs 2002). Two additional elements
should be also considered: the impact of new information technologies and
the rural-urban relationship.

Urban Planning

Effective urbanization requires urban planning to ensure basic infrastructure such as water, sanitation, power, public health, transport, and energy
systems to bring foreign investors, make markets work, and integrate urban economies into the global economy.

   *"Cities in Africa need, first and foremost, urban planners [rather than,
say, macroeconomists]."* (Sachs 2002)

In addition to the provision of infrastructure, urban planning should also take into account slum upgrading, the provision of land for new migrants to the city, and the existence of a legal instrument for land management and control.

Slum upgrading is by far preferable to slum clearance. Slum clearance will only transfer the same problem elsewhere. Slum upgrading, in conjunction with a set of complementary social and economic measures, can make a significant difference in the life of the poor. Experience has shown that the poor themselves are ready to participate actively in the process. Very often a tiny bit of help can spark positive change. Micro loans of as little as $50 have helped the waste-pickers of the Payatas landfill near Manila to secure loans for small businesses, land, and housing. The Zabbaleen society of waste-pickers in Cairo, Egypt have become organized and started a variety of income generating projects that involve composting and the recycling of rags and paper (O'Meara 2003). International organizations have identified the fundamentals of 'scaling up' slums. They include vigorous leadership and political will; strengthened government and voluntary institutions operating in tandem with clear policies, assigned roles, and cooperation; well managed, fiscally sound and organized city governments; ownership and full participation of the residents in the process of upgrading; and the provision of an appropriate package of affordable services (World Bank 2002). A good example of the concrete application of the aforementioned principles is the effort being made to combat urban poverty in Phnom Penh. The urban poor and their organizations are working with government agencies, NGOs, and international donors to develop homes and neighborhoods and generate forms of income (Asian Coalition for Housing Rights 2001).

A key factor for success is that upgrading should also deal with regulating the security of land tenure. Up to 80% of urban populations are living on land they do not own (UN-Habitat 2002c). Regulating tenure status removes a major source of economic and political security for households and for communities. It is a strong incentive for the active participation of local residents in improving their environment (World Bank 2003). The existence of legal instruments for land management and control is a necessary but not in itself sufficient condition to achieve better urban planning. There are too many examples of inadequate implementation and/or obsolete legislation not capable of covering new and rapidly changing situations. At the same time, its non-existence is a major hindrance to effective urban planning.

The majority of slums have been created as a result of rapid urbanization combined with the incapacity of existing urban areas to accommodate newcomers, either in public housing or through the private sector. Considering the rapidly increasing rates of urbanization and the influx of millions of new urban dwellers, it is also important for cities to develop contingency plans to face up to the challenge. A number of relevant ideas already exist (MIT 2001). It will be necessary to adapt them to the specific geographical and socioeconomic situation of the cities concerned. Particular emphasis will have to be given to medium-size cities, which are currently the fastest growing urban areas but inadequately equipped for growth.

Economic Development Strategy

An economic development strategy has to be tailored to the geographical specificities of the local area. For example when the special economic zones and the open port areas were created in China, coastal locations were reserved to create bases for export-led development. According to Jeffrey Sachs,

"... making an urban area an integrated part of the world economy is complementary to general macroeconomic stability, good governance and general good economics ... It requires a specific development strategy; for example, formulating export processing zones or creating industrial parks or science parks that have the characteristics that can attract international business to the local scene.... Tax holidays ... market mechanisms ... tenure rights for local land construction, mortgage market for urban construction of residential housing, microfinance for urban, small-scale entrepreneurship and urban housing finance." (Sachs 2002)

In this context, it is important to develop the appropriate conditions so that in addition to any Foreign Direct Investment, the endogenous private capital (which sometimes is considerable) can also be invested at the national/ local level. Cities are more likely to attract and keep such investment than rural areas. Employment creation is the crucial factor for the reduction of poverty, but there is often a 'Catch 22' situation.

"In Brazil, residents of poor favelas complained that employers would not hire anyone who has an address in favelas with a reputation for violence. Those favela residents in order to secure a job would give false addresses and even get fake electricity bills borrowed from friends in other locations." (Easterly 2002)

Urban Governance

The issue of the most appropriate and efficient urban governance is a very complex one. Many successful examples have been attributed to the granting of requisite autonomy to urban governments, such as in China or Brazil (Harris et al. 2001). Increased decentralization has often been seen as a panacea, which failed when it was not accompanied by a transfer of responsibilities, financial resources, and trained personnel. The role of the private sector is important. There are cities whose municipal authorities have a budget of one dollar per year per inhabitant. With such a budget the private sector is indispensable.

"With competent human resources, local authorities can improve the soundness and sustainability of their intervention programs and create favorable conditions for higher private investment." (Turchin et al. 2002)

Even today, the main actors of urban governance are considered to be the various local, social and environmental movements.

Effective urban governance should be seen as establishing the optimum equilibrium between various levels of government and actors, while considering the different and complementary roles of government at local, regional, national, and international levels. Governance consists of a network of bodies disposing of distinct responsibilities, the relative importance of which reflects the size and structure of urban areas, the composition of the population, the level of development of the society, and the strength of the economy. It has to ensure a balance between the direct local involvement in projects of their concern, on the one hand, and the efficient coordination at a higher level, on the other.

A successful local government will have to be able to develop five key dimensions to urban governance (Montgomery et al. 2003):

– **Capacity**—the ability of local governments to provide adequate public services to their citizens;

– **Financial**—the ability of local governments to raise and manage sufficient revenue;

– **Diversity**—the ability of government to cope with the extraordinary internal variation within cities and address the attendant issues of fragmentation and inequality;

– **Security**—the ability of government to deal with issues related to rising urban violence and crime;

– **Authority**—which is related to the increasing complexity of managing the jurisdictional mosaic as large cities grow and spread.

Information Technologies

The information revolution and the advent of new technologies could open up possibilities for individual and collective empowerment, information exchange, and knowledge accumulation that were previously unknown. Today, technology enables countries to enhance certain capacities almost instantaneously, with the wealth of experiences and expertise that can now be shared electronically. At the same time, there is a real danger that a new inequality is being added to the existing ones between developed and developing nations. Most developing countries are not actively taking part in the communications revolution. A new type of poverty looms, that of 'information poverty' (UN 1998).

The European Commission recognizes the positive role of Information Communication Technologies (ICTs) in development policy but limits their application to trade and development; regional integration and cooperation; support of macroeconomic policies and social services; transport; food security and rural development; and institutional capacity building (European Commission 2002c). The urban context in which such technologies have more chance to develop and have a greater impact is not included.

The introduction of new technologies and their appropriate and innovative use in developing countries in general and in urban areas in particular could be crucial in promoting sustainable urbanization in the short, medium, and long-term. Developing countries have a chance to bypass old technologies and leapfrog to using the latest technologies, for example in telecommunications and in intelligent traffic systems, in particular in view of the falling costs. Examples of poor countries leapfrogging to the technological frontier by imitating technologies from industrialized nations exist in a number of areas.

"Bangladeshi garment workers imitated Korean garment workers during their apprentice in Korea, and Bangladeshi managers imitated Korean managers. The result was a multibillion dollar garment export industry in Bangladesh." (Easterly 2002)

Access to technological innovations—more effective medicine, transportation, telephone, Internet—should not be seen as a result of economic growth and income increase but as a tool towards growth and development. Investments in technology and the relevant education can equip people with better tools to make them more productive and prosperous (UNDP 2001). Pricing is crucial.

Currently, most examples of pilot applications of information technology in less developed countries are located in rural areas. They refer to the use of mobile phones and Internet centers (telecenters) and virtual telephones connecting remote rural villages (Caspary 2002). There could be even more promising avenues to be made in urban areas. In developed countries, information and telecommunications technologies are shaping the form and function of cities in a more complex way than the automobile did in earlier eras (New York University 2001). They could have a positive impact on urban areas in developing countries. Today, ICT infrastructures and applications in cities of developing countries follow a model that reinforces the urban divide instead of diffusing the benefits to the totality of the urban area. ICTs concentrate in enclaves reserved for the rich or for foreign investors (UNCHS 2001a). Efforts should be made to exploit ICT capabilities to support development models that are more equitable, democratic and sustainable (ibid).

There are many examples of urban applications of information technologies. More and more cities set up wireless networks in the US and provide free wireless Internet access in order to boost economic development and make them more attractive (Markoff 2003). In Rwanda, there are no legal instruments for the control and management of land. Kigali's rapid urbanization has been anarchic with the majority of land being occupied by illegal settlements. Putting order into chaos can not be done with the 'classical' methods of cadastre. However, new GIS technologies could achieve much more rapid and efficient results.[15] Many of the world's mega-cities in developing countries need or will soon need advanced traffic management systems, such as the one recently introduced in London to control the flow of cars through their city centers (Financial Times 2003).

Urban-Rural Relationship

Support for cities will not necessarily increase rural to urban migration. The latter will take place whatever the context. The high density of rural areas in developing countries combined with the population explosion and an increased rural productivity will lead to a great number of rural inhabitants to migrate to cities whatever the situation of the receiving urban areas.[16] Improved urban conditions and city economic development will have a beneficial spillover to the rural population. The reverse is less probable. Urban proximity has always been, with the right support, a positive factor for economic development. With the present rates of urban population

growth in the developing world, if no action is taken at the urban level, the
ensuing disaster will have a direct impact on rural areas and, consequently,
on the national economy. The urban-rural relationship has been recently
the subject of a number of studies. It is becoming widely recognized that
successful rural development rests on the economic vitality and sustainability of small and medium-size urban centers. Paying more attention to
the urban issues will make it easier to break the artificial division between
urban and rural development (Satterthwaite and Tacoli 2003). A closer
look at the urban-rural relationship is necessary today even more than before. This relationship is expressed in two ways. On the one hand, there is
the "complex world with both urban and rural characteristics that clings to
the urban conglomerations" (Griffon 2002) and, on the other hand, the
world of small and intermediate urban centers which have been the fastest
growing settlements in the developing world.

In what has been termed as 'rurban' areas, we have to deal with the
'worst' of both worlds: densely populated areas with poor housing conditions "rubbing shoulders with farming and cattle breeding," leading to all
sorts of problems: social conflict, pollution, and poverty (ibid). Findings of
a recent report have shown that on a number of issues the urban poor demonstrate behavior patterns very similar to those of rural populations
(Montgomery et al. 2003).[17] However, the integration of rural population
into the system of a large urban agglomeration is not an easy task. It requires innovative measures and planning. Of particular importance is the
development of appropriate measures to safeguard, rationalize and promote urban agriculture, which is both a provider of employment as well as
a critical food-security valve for poor urban households (UNCHS 2001a).[18]
UNDP estimates that 800 million people are engaged in urban agriculture
worldwide today, the majority in Asian cities (ibid). Of these, 200 million
are considered to be market producers, employing 150 million people full-
time. Authorities often try to prevent urban food production, fearing the
risk of contamination, rather than finding solutions for it (Argenti 2002).
Innovative approaches are required to expand production and prevent possible risks.

The small and intermediate urban centers in low and middle-income nations are expected to account for over 40% of the increase in the world's
urban population between 2000 and 2015 (Tacoli 2003). Their main problem is the serious lack of infrastructure and services. More often that not
they have been excluded from mainstream development programs. The
trend of rural to urban migration cannot be stopped. Urban centers are crucial for regional rural development both in order to absorb population
growth and provide services for the surrounding rural area.

Conclusions

A one dimensional rural and/or sectoral approach to development might have been considered adequate up to now, but it has not produced any spectacular results. It is difficult to justify its continuation in the future. Such an approach runs counter to the global trends of rapid urbanization, impoverishment of urban populations and the complexity of the problems involved. It is not in conformity to the existing and/or emerging policy and scientific paradigms, which promote sustainability, integration, and local empowerment. Many western societies are implementing these principles in their territories. Examples include the EU Urban Integrated Approach (the URBAN initiative)[19] and US Smart Growth for urban areas.[20]

There is not one single remedy appropriate for all urban areas. There are urban areas with high potential that have not succeeded and others with lower potential that have. Geography plays also a crucial role. It is not the same to be located in a coastal area or in the interior of a country or a continent. It is also clear that the global macroeconomic environment, the way major international players handle trade and development issues together with political stability, are essential factors in changing the condition of the majority of urban populations in less developed countries.

Notwithstanding the importance of the overall global context, priority should be also given to tackle the major challenges stemming from the unprecedented rates of urbanization in the developing world. Current trends show that many cities in these countries fail to live up to their potential as motors of economic and social development. Although we have acquired a sufficient understanding of the problems, the best ways of tackling them, and the constraints that impede their implementation, this knowledge has not yet been translated into concrete action.

A key factor in this process is the role and the policies of international donors. It is time to elaborate strategies that will include support for an increased number of *Integrated Urban Programs*. This might not necessarily mean an increase of available resources, (which would be unrealistic under the present economic and political conditions), nor even a change in the balance between the resources available to rural and urban areas, although this might be necessary in the future. What is more realistic is a gradual shift of the emphasis from sectoral to more integrated approaches. The successful implementation of an integrated approach will also require an increased effort to coordinate the activities of various international donors.

Endnotes

[1] The views expressed by Marios Camhis, an official of the European Commission, are his own and do not necessarily reflect the Commission's position.

[2] Ecological Footprint is a function of population size, average per capita consumption of resources and the resource intensity of the technology used. See www.unep.org/geo/geo3.

[3] Op cit p. 233

[4] english.peopledaily.com.cn/199912/09/eng19991209X101.html

[5] www.worldbank.org/urban/poverty

[6] www.fao.org/ag/magazine/0206sp2.htm

[7] www.worldbank.org/urban/upgrading

[8] europa.eu.int/comm/development/body/legislation/docs/council_statement.pdf#zoom = 100

[9] www.bestpractices.org

[10] www.worldbank.org/urban/agenda.htm

[11] www.wds.worldbank.org/servlet/

[12] CDS were first proposed by the World Bank in 1997 to cover poverty, economic development, the environment, city management and finance.

[13] Information based on interview with USAID

[14] europa.eu.int/comm/europaid/projects/asia-urbs/documents/strategy_eu_asia_2001.pdf

[15] Presentation by a Rwanda official at a seminar at the Woodrow Wilson International Center for Scholars, May 2003

[16] www.worldbank.org/data/wdi2002/environment.pdf

[17] Fertility rates, use of contraceptives, levels of reproductive health, etc.

[18] www.uneporg/geo/geo3/english/048.htm#fig9

[19] The principles of the URBAN initiative include targeting on small areas of severe deprivation, focusing on issues of community interest (social inclusion, environment, employment), and promoting local partnership.

[20] Smart Growth initiative promotes mixing of land uses, creation of a range of housing opportunities, transportation choices, and community and shareholder collaboration in development decisions.

References

Argenti O (2002) Urban Food Safety. *FAO Magazine,* June 2002. www.fao.org/ag/magazine/0206sp2.htm

Asian Coalition for Housing Rights (2001) Building an urban poor people's movement in Phnom Penh. *Environment & Urbanization,* Vol. 13, No. 2

Atkinson A (2002) International Co-operation in pursuit of sustainable cities. In: Westendorff D, Eade D (eds) *Development and Cities.* Oxfam GB, Oxford

Caspary G (2002) *Information Technologies to Serve the Poor.* www.dse.de/zeitschr/de102-3.htm

Chatterjee G (2002) Consensus versus Confrontation. *Habitat Debate* June 2002, Vol. 8, No. 2. UN, Nairobi

Cities Alliance (2000*) Local Partnership: Moving to Scale.* Report of the Meeting of the Cities Alliance Public Policy Forum. www.citiesalliance.org

Cities Alliance (2002) *Annual Report.* www.citiesalliance.org

Connolly P (2001) Mexico City: Our Common Future? In: Harris Jonathan M. et al. (eds) *A Survey of Sustainable Development.* Island Press, Washington DC

Easterly W (2002) *The Elusive Quest for Growth.* MIT Press, Cambridge MA

European Commission (2000a) *URBAN Success Stories: Building a better tomorrow in deprived neighborhoods.* Office of Publications, Luxembourg

European Commission (2000b) *Development Policy Statement.* /europa.eu.int/comm/development/body/legislation/docs/council_statement.pdf#zoom = 100

European Commission (2001) *Fighting Poverty.* Office of Publications, Luxembourg

European Commission (2001) *Information and Communication Technologies in Development: The role of ICTs in EC development policy.* COM/2001/770 final

European Commission (2002a) *Europe's Response to World Ageing.* COM/2002/0143 final

European Commission (2002b) *Fighting Rural Poverty.* COM/2002/0429 final

European Commission (2003) *World Energy, Technology and Climate Policy Outlook 2030.* DG Research, Luxemburg

European Commission (2005) *EU Strategy for Africa: Towards a Euro-African pact to accelerate Africa's development.* COM/2005/ 489 final

European Environment Agency (2003) EU Greenhouse emissions rise for second year running. Press Release, May 2003. org.eea.eu.int

Fahn J (2003) *A Land of Fire: The Environmental Consequences of the Southeast Asian Boom.* Westview Press, Boulder CO

FAO (2001) *Ethical Issues in Food and Agriculture.* www.fao.org/DOC-REP/003/X9601E/x9601e02.htm

Financial Times (2003) Traffic congestion charges. *Financial Times,* 13 February 2003

Garrett J, Ruel M (1999) IFPRI. *Choices,* December 1999

Griffon M (2002) Sustainable Development of "Rurban" Areas is Crucial. *The Courier ACP-EU*, No. 195, November-December 2002

Harris JM et al. eds (2001) *A Survey of Sustainable Development.* Island Press, Washington DC, p. 127

Harris N (2002) Cities as Economic Development Tools. *Urban Brief,* December 2002, Woodrow Wilson International Center for Scholars

Hicks JF (1998) *Africa Region.* World Bank at the International Conference on Research Community for the Habitat Agenda Forum of Researchers on Human Settlements Geneva, July 6–8, 1998

ICMA, Smart Growth Network (2003) *Getting to Smart Growth: 100 Polices for Implementation.* Smartgrowth.org

IFPRI (2000) *The Urban Poor: Complex Problems Face Developing Countries as Urbanization Rises.* www.ifpri.org/reports/00spring/00sprf.htm

Leitmann J (2003) Urbanization and Environmental Change: Issues and options for human security. *AVISO*, April 2003, No. 11. GECHS, Ottawa

Lvovsky K (2001) *Health and Environment.* IBRD/WB, Washington DC, p. 37

Markoff J (2003) More cities set up wireless networks. *New York Times,* 6 January 2003

MIT (2001) *What is Urban Upgrading, Issues and Tools, Spatial Planning Grids for Directing Growth.* web.mit.edu/urbanupgrading

Montgomery MR, Stren R, Cohen B, Reed HE eds (2003) *Cities Transformed: Demographic Change and Its Implications in the Developing World.* The National Academies Press, Washington DC

New York University (2001) *Urban Research Institute.* www.informationcity.org

OECD (2001) *Strategies for Sustainable Development: Practical Guidance for Development Co-operation.* webnet1.oecd.org/EN/countrylist

O'Meara SM (2003) Uniting Divided Cities. In: *State of the World 2003.* A Worldwatch Institute Report, pp. 140–141, W. W. Norton & Co, New York, London. www.worldwatch.org/pubs/sow/2003/

Recife Declaration (1996) *Recife International Meeting on Urban Poverty,* Recife, Brazil, 17–21 March 1996. www.unhabitat.org/programmes/ifup/rde.asp

Sachs JD (2002) *Inaugural Address,* Urban Research Symposium, 9 December 2002. World Bank, Washington DC

Sachs JD (2005) *Investing in Development, A Practical Plan to Achieve the Millennium Development Goals.* UN Millennium Project, New York

Satterthwaite D, Tacoli C (2003) *Rural-Urban Transformations and the Links between Urban and Rural Development.* Paper in Workshop, Integrating Rural Development and Small Urban Centers. World Bank, Washington DC

Stephens C (2002) Urban Health in the Twenty-First Century: Challenges of Privatization, Participation, Individualism, and Citizenry. In: Turchin J et al. eds *Democratic Governance and Urban Sustainability.* Woodrow Wilson International Center for Scholars, Washington DC pp. 77–88

Sundt N (1999) *Chinese emissions still growing.* www.globalchange.org/dataall/98dec9a.htm

Tacoli C (2003) *The potential and actual role of small and intermediate urban centers in rural regional development and poverty reduction.* Paper in Workshop, Integrating Rural Development and Small Urban Centers, March 2003. World Bank, Washington DC

Turchin J, Varat DV, Ruble BA eds (2002) *Democratic Governance and Urban Sustainability.* Woodrow Wilson International Center for Scholars, Washington DC

UN (1998) United Nations Administrative Committee on Coordination (ACC), Statement on Universal Access to Basic Communication and Information Services, April 1997. Quoted in ITU: *World Telecommunication Development Report* 1998, p. 10

UN (2000) General Assembly, *United Nations Millennium Declaration,* 8 September 2000. United Nations, New York

UN (2002a) *International Migration Report 2002.* United Nations, New York

UN (2002b) *World Summit on Sustainable Development, Plan of Implementation.* www.johannesburgsummit.org

UN (2002c) *Report of the International Conference on Financing for Development.* ods-dds-ny.un.org/doc

UNCHS (2001a) *Cities in a Globalizing World.* Earthscan Publications, London

UNCHS (2001b) *The State of the World's Cities.* UNCHS Habitat, New York

UNDP (2001) *Human Development Report 2001: Making new technologies work for human development.* Oxford University Press, New York

UN Millennium Project (2004) *Interim report of the taskforce* 8, February 2004. www.unmillenniumproject.org

UN Population Division (2001a) *World Urbanization Prospects: The 2001 Revision.* United Nations, New York

UN Population Division (2001b) *The State of World Population.* www.unfpa.org

UN-Habitat (2002a) *Sustainable Urbanization Bridging the Green and Brown Agendas.* DPU, London

UN-Habitat (2002b) *Sustainable Urbanization.* UN-Habitat, Nairobi

UN-Habitat (2002c) *Press Conference on Sustainable Urbanization.* www.un.org/ events/wssd/pressconf/030827conf2.htm

UN-Habitat (2003) *Water and Sanitation in the World's Cities: Local Action for Global Goals.* www.earthscan.co.uk

USAID (2003) *Millennium Challenge Account.* www.USaid.gov/mca/

Westendorff D, Eade D eds (2002) *Development and Cities.* Oxfam GB, Oxford

White House (2002) *President announces clear skies and global climate change initiatives.* www.whitehouse.gov/news/releases

Wilson S (2002) Urban Anti-Rebel Raid a New Turn in Colombian War. *Washington Post* Oct 24, 2002

Wolfensohn JD (2003) Towards a more secure world. Remarks at the Institute of Foreign Affairs. web.worldbank.org/WBSITE/EXTERNAL/NEWS/

World Bank (1999) *Comprehensive Development Framework.* www.worldbank.org/cdf/

World Bank (2001) *Urban Services to the Poor Thematic Group* www.worldbank.org/urban/upgrading/slum.html

World Bank (2002) *Brazil Progressive Low-Income Housing—Alternatives for the Poor.* Report No. 22032-BR

World Bank (2003) *World Development Report 2003: Sustainable Development in a Dynamic World,* World Bank and Oxford University Press, Washington DC

Zwingle E (2002) Megacities. *National Geographic Magazine,* November, pp. 70–99

Websites:

english.peopledaily.com.cn/199912/09/eng19991209X101.html

europa.eu.int/comm/development/

www.bestpractices.org

www.fao.org/ag/magazine/0206sp2.htm

www.uneporg/geo/geo3
www.worldbank.org/data/wdi2002/environment.htm
www.worldbank.org/urban/poverty
www.worldbank.org/urban/upgrading
www.worldwatch.org

Sustainable Economies—Local or Global?

HELENA NORBERG-HODGE

Any attempt to discuss the future of sustainable development needs to include an analysis of the impact of the deregulation of trade, or economic globalization. Though centuries old, this process has gained momentum over the last few decades. The deregulation of international trade has propelled businesses to merge into ever larger transnational monopolies. These giant transnational corporations (TNCs) in turn have gained more and more power over governments and society as a whole, shaping policy, research priorities, and the dissemination of information in academia and the media. They have also transformed the whole debate about sustainability. From the outset, the author would like to make it clear that this paper seeks to look at the issues or structures, not to demonize individuals or corporations. However, the concentration of power through trade deregulation is a vitally important reality that needs to be discussed widely and openly. In so doing, it will become apparent that sustainable development requires a shift in direction—from globalizing economic activity towards localizing it.

Economic globalization is not a natural process; it is based on very specific policies. In particular, a range of international treaties, from the GATT after the Second World War to more recent rounds of negotiations, has liberalized the global movement of goods and capital. This has led to increasing economic concentration in the hands of those banks and businesses that operate at the global level. As part of this process, a northern model, based on industrial production, trade, and economic growth is being systematically imposed globally. Proponents claim that this is the only way to create employment and raise standards of living in rich and poor nations alike. The deregulation of international trade and investment lies at the heart of this process. This, in turn, means a dramatic increase in the transportation of goods across the world. Nowadays, almost everything we use, from building materials, clothes, food and drink, comes from thousands of miles away and this trade needs massive infrastructures—long-distance

99

M. Keiner (ed.), The Future of Sustainability, 99–115.
© 2006 *Springer. Printed in the Netherlands.*

transport, huge centralized energy plants and high speed communications—to support it. One needs only to look at the mounting environmental and social crises of the world today, from global warming and loss of biodiversity to dwindling oil supplies, to see how this economic model cannot be sustained, and how it is bringing a tidal wave of destruction in its wake. Globalization and sustainability are simply incompatible.

In a sense, the countries of the South have subsidized today's globalized economy for the past 500 years, at great expense to their own cultures, their land and their economies. The current dominance of the western industrial model could never have arisen without prolonged access to the South's raw materials, labor (including slave labor), and markets.

Although it is generally believed that the infamous era of conquest and colonialism is behind us, today's 'development', 'structural adjustment' and 'free trade' are simply new forms of the same exploitative process. In its present phase—economic globalization—policymakers are pushing the western industrial system into the farthest corners of the planet, attempting to absorb every local, regional and national economy into a single centrally managed world economy based on ever increasing trade.

Trade between peoples and nations is nothing new—it is an activity people have engaged in for millennia. But in the past, trade for most societies was nearly always a secondary concern, while the primary economic goal was meeting people's needs and wants using the resources available within relatively short distances. Only once essential needs had been met locally did questions of trading surplus production with outsiders arise.

In the modern era, however, trade has come to be pursued as an end in itself. This emphasis can be traced to an 1817 theory of political economist David Ricardo, which holds that nations are better off if they specialize their production in areas where they excel—those in which they hold a 'comparative advantage' in relation to other countries—and then trade their surpluses for goods they require but no longer produce. The ostensible goal is to increase 'efficiency', but the result has been a system that is highly *inefficient* and wasteful. This theoretical model does not take into account the real costs of increased trade. Since most of these costs are paid by taxpayers through subsidies or 'externalized' to the public or the environment, the theory's shortcomings are not immediately apparent. Comparative advantage still guides government planning and decision-making today and is at the heart of the dogma of 'free trade'.

In thrall to an outdated economic theory, governments are making massive investments in trade-based infrastructures, signing trade treaties that open their economies to outside investment, and scrapping laws and regulations designed to protect national and local businesses, jobs and resources.

In many ways, national sovereignty is being relinquished to undemocratic supranational bodies like the World Trade Organization (WTO), in the mistaken belief that trade is always good and that more trade is always better.

The result of these policies has been an explosive growth in international trade, which has multiplied twelve-fold since the 1940s—almost two-and-a-half times faster than the growth in output. Imports and exports now make up a much larger proportion of economic activity than ever before, with traded goods totaling some US $5.5 trillion annually (IMF Statistic Department 1999).

Whole economies are becoming dependent on trade, and virtually every sphere of life is being affected. The impact on food, one of the only products that people everywhere need on a daily basis, is particularly revealing. Today, one can find apples shipped from New Zealand in apple-growing regions of Europe and North America; kiwis from California, in turn, have invaded the shops of New Zealand. In Mongolia, a country with 10 times as many milk-producing animals as people, shops carry more European dairy products than local ones. England imports more than 100,000 tonnes of milk each year, then turns around and exports roughly the same amount. In much of the industrial world, the average plate of food travels thousands of miles before reaching the dinner table.

What are the benefits of transporting basic foods such distances, when they can be (and indeed for centuries have been) produced locally? How can these arrangements be described as economically 'efficient' and sustainable? As we will see, this excessive trade is disastrous; even the wealthiest suffer from the ensuing stress, pollution and social breakdown.

Hidden Subsidies

Proponents of globalization point to the lower cost of many traded goods as proof that economic efficiency is at work. However, a close look at the way the global economy is subsidized deflates this argument. Not only do governments promote trade through international treaties, they do so by handing out direct subsidies to the trading sector of their economies. In the US, for example, The Market Access Programme devoted $100 million in 2002 (which will increase to $200 million by 2006) to companies like Sunkist, Miller Beer, Campbell's Soup, McDonald's, and M&M Mars to advertise their products abroad.[1]

Perhaps more importantly, governments also subsidize the global economy *indirectly* through investments in the infrastructures a trade-based economy requires. These taxpayer-funded infrastructures include the following:

- **Long-distance transport networks**—multilane highways and motorway networks, shipping terminals, airports, high-speed rail, container facilities, etc.;

- **Energy infrastructures**—large, centralized electric power plants (including nuclear power stations and hydroelectric dams), petroleum facilities, gas and coal slurry pipelines, etc.;

- **High-speed communications and information networks**—the Internet, satellites, telephone networks, television, and radio;

- **Research and development institutions**—facilities that develop labor-displacing technologies for industry and agriculture and technologies to expand and modernize the physical infrastructures supporting the global economy.

One effect of this system of direct and indirect subsidies is that the price of goods transported halfway around the world can seem artificially 'cheap' in comparison to goods produced next door. Ignoring or externalizing environmental costs has a similar effect. Thus, garlic transported to Spain all the way from China can be half the price of locally grown garlic mainly because neither the pollution involved in its transport nor the cost of the transport infrastructure itself are reflected in its price.

The Big Get Bigger

Subsidized trade-based infrastructures have helped to expand the size of markets to global proportions. In the process they have enabled large-scale, globally oriented corporations to invade and absorb the markets of small-scale, locally oriented enterprises. Though they are unaccountable to any electorate, many of these corporations are now so large that they wield more economic and political power than national governments: at least half of the 100 largest economies in the world are, in fact, not countries but corporations.[2] Five hundred corporations now control 70 percent of total world trade. In 2002, a report showed that five privately owned companies (Cargill, Continental, Louis Dreyfus, Andre, and Bunge) control up to 90 percent of global grain trade (Murphy 2002).

'Free trade' treaties like NAFTA and GATT were designed to give those corporations freer reign by forcing countries to remove any tariffs or regulations that could be seen as barriers to trade. Proponents of 'free trade' insist that corporations should have the right to invade any and every market; already many measures adopted by national governments taken in the public interest have already been construed as trade barriers by the WTO and have been struck down.

Nonetheless, governments naively support what they think of as 'their' multinationals, even though in today's competitive market, corporations simply cannot maintain loyalty to place. Tax breaks, capital grants, free land, and lax environmental and worker safety rules can easily lure a corporation away from its country or region of origin; equal or greater handouts must be offered to induce them to stay where they are. A typical company providing several hundred jobs can expect to be provided capital grants for its building and machinery costs, low interest loans, subsidized training for its new labor force, and a host of tax relief measures. Small local businesses, given no such subsidies, cannot hope to survive this unfair competition. This process has contributed greatly to the loss of vitality of entire communities and has triggered an international 'race to the bottom', with social, environmental and health standards in all countries heading towards the lowest common denominator.

If there is anything free about 'free trade', it is the freedom it confers on corporations to move their operations to countries where taxes and labor costs are low, environmental regulations are weak, and taxpayer-funded subsidies are large.

Squeezing out the Small

Structurally, the globalizing economy systematically favors footloose corporations over small, rooted businesses. Long-distance transport networks, for example, make it possible for huge agribusinesses and corporate marketers to deliver their products worldwide, helping them absorb the markets of businesses selling locally produced goods. Publicly funded global communications networks are of little use to the local family farmer or the corner grocer, but they enable transnational corporations to wield centralized control over their widely dispersed activities and transfer capital around the world at the stroke of a computer key. And while small local enterprises thrive by filling the numerous economic niches cultural diversity provides, transnational corporations depend on markets that have been homogenized—in part by advertising campaigns conducted through global media. As they grow in economic power, the sheer size and financial power of transnational corporations enables them to extract price breaks from suppliers and lending institutions, as well as concessions from governments and regulatory bodies.

On such a tilted playing field, how can a local grocer possibly hope to compete with a large supermarket chain? How can sustainable family farmers survive when pitted against heavily subsidized corporate agribusinesses? And how can local retailers compete with huge mega-stores and on-line businesses? Is it any surprise that with each year the number of independent businesses, shopkeepers, and small farmers continues to plummet?

The True Costs

The globalized economy has not only been disastrous for the small shopkeeper and the family farmer: economic globalization is affecting us all—as individuals, families and communities—and is putting the biosphere under increasing strain. More specifically, globalization is leading to significant global crises:

– **Erosion of democracy:** As decision-making becomes centralized into unelected, unaccountable bodies like the World Trade Organization (WTO), the International Monetary Fund (IMF), and the European Commission, the influence of the individual steadily shrinks—even in nominally democratic countries. People may still have the right to vote for national and local leaders, but as political institutions, both left and right, adopt identical policy measures influenced by and reflecting the wishes of mobile corporations, voting can become all but meaningless.

– **Loss of government autonomy:** As dependence on the global economy grows, it becomes increasingly difficult for even nation-states to protect their citizenry or environment from the dictates of international finance and transnational capital. In the South (and increasingly in the North as well) governments themselves are losing autonomy, as they are forced to reshape their economies to fit the contours of the global economy. Countries are encouraged to make their production more 'efficient' by focusing on just two or three key commodities for the global marketplace. This means building up the infrastructures necessary for export-oriented industry and agriculture, which requires tremendous amounts of capital. This capital must be borrowed, and if global demand for exports declines, countries may be unable to repay their loans, forcing them further into debt. They are then pressured to undertake 'structural adjustment' programs to further enhance international 'competitiveness'. This means cutting back on social spending, limiting restrictions on investment, and providing still more funding for infrastructures.

World Bank/IMF lending to Southern countries is typically made conditional on such programs, and indeed the vast majority of these countries have been subjected to them. The continual loan repayments, for which the interest alone may be equal to a large percentage of the country's annual budget, require surpluses that can only be generated by trading away natural resources or a significant portion of national output. In this way, entire nations are not only impoverished by a vicious debt cycle, they are also ensnared into ever greater dependence on the global economy.

- **Economic destabilization:** Tied to a complex system of imports and exports, countries are becoming more and more tightly linked to a volatile global economy over which they have no control. Natural disasters, wars, and economic slumps in one part of the world can have a direct impact on countries many thousands of miles away. American farmers, for example, found no market for half their grain harvest in 1999, thanks to the financial crisis that struck Asia—a market on which those farmers had become dependent (Weiner 1999, Johnson 1999).

  The speculative nature of most global investment makes the entire system even more unstable. In fact, the most traded product on global markets today is not something you can clothe or feed yourself with—it is money. Every day of the year, roughly $1.3 trillion is gambled on international currency markets, 30 times more than the daily GDP of all the developed countries combined (Lietaer 1997). More than 95 percent of this involves pure speculation, leading many experts to conclude that the system is so unstable its eventual breakdown is assured: 'It is only a question of when', argues international financier George Soros (Morris 1997). A small sample of that breakdown occurred in 1998, when unfettered speculation in the currencies of Southeast Asia led to financial crisis and recession across the region, with severe economic repercussions felt worldwide.

- **Urbanization:** Industrial economic growth erodes rural economies so that only 2% of the population remains on the land in highly industrialized countries. Globalization is accelerating that trend, leading to a massive population shift from rural areas to the cities. This is particularly true in the South, where the growth economy is steadily breaking down more self-reliant systems, leaving people little alternative but to migrate to ever expanding cities. Even in the most industrialized countries, the urbanizing process continues: jobs in the global economy are concentrated in sprawling metropolitan areas and their suburbs, while rural regions are systematically sapped of economic vitality.

This unhealthy urbanization not only hollows out rural communities, it also leads to a host of urban problems: overcrowded slums (particularly in the South), loneliness, alienation, family breakup, poverty, crime, and violence. Urbanization also contributes to a massive increase in resource use and pollution: virtually every material need of urbanized populations must be shipped in from elsewhere, while the resulting waste— much of which would be of use in a rural setting—becomes a highly concentrated source of pollution. If the urbanization of the world's population continues at present rates, it will only lead to further social and environmental breakdown.

– **Loss of food security:** The heavy emphasis on exports has led to a rapid decrease in agricultural diversity, with thousands of local varieties abandoned for the relative few suited to monocultural production and favored by short-term economic trends. Overall, approximately 75 percent of the world's agricultural diversity has been lost in the last century, a narrowing of the genetic base that threatens food security everywhere.[3]

Increasing control by a handful of corporations over the world's food supply also threatens people's access to food, particularly those without enough money to meet corporate profit expectations. Today, in fact, when food is more tightly controlled by corporations than ever before some 840 million people are undernourished (FAO 2002), even though more than enough food is produced to adequately feed everyone on the planet.

– **Growing gap between the 'haves' and the 'have-nots':** Economic globalization is leading to a widening gap between rich and poor, both between the countries of the North and the South and within individual countries themselves. Already the wealth of 350 billionaires equals the annual income of the poorest 45 percent of the world's population, and yet, the inequity continues to grow worse. The situation is exacerbated by the mobility of transnational corporations and capital, which operates to drive down wages everywhere. Production for global markets, meanwhile, is increasingly dependent on large-scale computerized and automated processes, thereby further marginalizing human labor. If much of the world's population is to continue leaving their villages in search of scarce jobs in the cities, how will the majority survive, jobless and with little prospect of future employment?

– **Increased ethnic and racial conflict:** Globalization is replacing the earth's cultural diversity with a uniform Western monoculture. Historically, the erosion of indigenous cultural integrity was a conscious goal of the architects of colonialism: Colonial officers were advised to "deliberately tamper with the equilibrium of the traditional culture so that change will become imperative."[4] This process continues into the present day, both as conscious policy and as a result of the insidious effects of global media and advertising.

Every day, people around the world are bombarded with media images that present the modern, Western consumer lifestyle as the ideal, while implicitly denigrating local traditions and land-based ways of life. The message is that the urban is sophisticated and the rural is backward; that imports of processed food and manufactured goods are superior to local products; that 'imported is good, local is crap', in the words of an advertising executive in China.[5]

People are not only being lured to abandon local foods for McDonald's hamburgers and local dress for designer jeans, they are induced to remake their own identities to emulate the glamorous blonde-haired, blue-eyed stars of Baywatch or Dallas. For the vast majority around the world, the attempt to live up to this artificial ideal will prove impossible. What follows is often a profound sense of failure, inferiority and self-rejection. When combined with the cultural uprooting, poverty and hopelessness that permeates much of the 'developing' world, the predictable outcome is a rise in fundamentalism, ethnic conflict, and violence.

– **Environmental breakdown:** Globalization is intensifying the already serious ecological consequences of industrialization. Despite western faith in the ability of high technology and human ingenuity to solve these problems, we have already exceeded the biosphere's capacity to absorb the impact of industrial activities. The soil upon which food production ultimately depends is being rapidly lost thanks to industrial farm practices. The global timber industry, land speculators, and oil and mining industries have decimated whole tracts of irreplaceable forest. Our air and water are increasingly polluted, and mountains of toxic waste and nuclear debris continue to grow. The introduction of genetically modified crops poses the threat of irreversible 'genetic pollution'. The planet's immense diversity of plant and animal species is being eroded at the rate of at least 50,000 species annually, ranking this as one of the planet's great extinction waves (Wilson 1992). Perhaps worst of all, deforestation, ozone depletion, and greenhouse gas emissions are making global weather patterns more extreme, unpredictable, and violent.

Accelerating these trends through globalization is simply incompatible with environmental sustainability: more trade means more transport, which means more pollution and CO_2 emissions; the consolidation and 'modernization' of agriculture means more soil erosion, more toxic agro-chemicals, and more resource-intensive urbanization; the continued building of transport infrastructures and extraction of fossil fuels means more destruction of habitats and loss of biodiversity. Clearly, this finite planet does not have the capacity to sustain an economic system based on unlimited growth. Yet the premise of globalization is that more of the world's people—all of them, in fact—should be encouraged to enlist in this destructive system.

Ultimately, today's increasingly globalized economy has no winners. Workers around the world are left either unemployed or in low-paying jobs with minimal safety conditions and little job security. Millions of small and medium-sized businesses are closing down, as transnational corporations take over markets of every kind. Small farmers are devalued, financially destroyed, and drawn off to the mega-cities, leaving behind villages and small towns devoid of economic and cultural vitality. The environment is becoming increasingly polluted and destabilized. In the long run, not even the wealthy few can escape these problems; they too must survive on an ecologically degraded planet, and suffer the consequences of a social fabric ripped apart.

Localization

If globalizing economic activity leads to such unsustainable and destructive development, *localizing* would seem an obvious step towards sustainability. Localization does not mean encouraging every community to be entirely self-reliant; it simply means shortening the distance between producers and consumers wherever possible, and striking a healthier balance between trade and local production. Localization does not mean that everyone must go 'back to the land', but that the forces now causing rapid urbanization should cease. Localization does not mean that people in cold climates should be denied oranges or avocados, but that their wheat, rice or milk—in short, their basic food needs—should not travel thousands of miles when they could all be produced within a fifty-mile radius. Rather than ending all trade, steps towards localization would aim at reducing unnecessary transport while encouraging changes to strengthen and diversify economies at the community as well as national level. The degree of diversification, the goods produced, and the amount of trade would naturally vary from region to region.

Reversing our headlong rush towards globalization would have benefits on a number of levels. Rural economies in both North and South would be revitalized, helping to stem the unhealthy tide of urbanization. Farmers would be growing for local and regional rather than global markets, allowing them to choose varieties in tune with local conditions and local tastes, thus allowing agricultural diversity to rebound. Production processes would be far smaller in scale, and therefore less stressful to the environment. Unnecessary transport would be minimized, and so the greenhouse gas and pollution toll would decrease, as would the ecological costs of energy extraction. People would no longer be forced to conform to the impossible ideals of a global consumer monoculture, thereby lessening the psychological pressures that often lead to ethnic conflict and violence. Ending the manic pursuit of trade would reduce the economic and hence political power of TNCs, and eliminate the need to hand power such supranational institutions as the WTO, thereby helping to reverse the erosion of democracy.

Rethinking Basic Assumptions

In these and many other ways, a shift towards the local simply makes sense. Nonetheless, calling for this shift in direction tends to elicit a chorus of objections. Some claim that the promotion of decentralization is 'social engineering', involving serious dislocations in the lives of many people. While it is true that some disruption would inevitably accompany a shift toward the local, it is far *less* than is already resulting from the current rush towards globalization. It is, in fact, today's 'jobless growth' society that entails social and environmental engineering on an unprecedented scale: vast stretches of the planet and entire economies being remade to conform to the needs of global growth, just as people around the world are being encouraged to abandon their languages, their foods, and their architectural styles for a standardized monoculture.

Another objection is the belief that people in countries of the South need Northern markets in a globalized economy to lift themselves out of poverty, and that a greater degree of self-reliance in the North would therefore undermine the economies of the Third World. In large measure, this view arises from the erroneous belief that poverty in the South is simply due to a lack of development—that this is how people lived before they had benefited from western-style modernization. However, an honest appraisal of the historical record, beginning in the *pre*-colonial era, reveals that Third World poverty is primarily a *consequence* of development, from the colonial period to the present day.

The truth of the matter is that a gradual shift towards smaller scale and more localized production would benefit both North *and* South, and would facilitate meaningful work and more employment everywhere. The globalized economy requires the South to send a large portion of its natural resources to the North as raw materials; its best agricultural land must be devoted to growing food, fibers, and even flowers for the North; and a good deal of the South's labor is employed in the cheap manufacture of goods for Northern markets. Rather than further impoverishing the South, producing more *ourselves* would allow the South to keep more of its own resources, labor and production for *itself*. Globalization means pulling millions of people away from sure subsistence in a land-based economy into urban slums from which they have little hope of ever escaping. Diversifying and localizing economic activity offers the majority, in both North and South, far better prospects.

The idea of localization also runs counter to today's general belief that fast-paced urban areas are the locus of 'real' culture, while small, local communities are isolated backwaters, relics of a past when small-mindedness and prejudice were the norm. It is assumed that the past was always brutish, a time when exploitation was fierce, intolerance rampant, and violence commonplace—a situation that the modern world has largely risen above. These assumptions echo the elitist or racist belief that modernized people are superior, even more highly evolved than their 'underdeveloped' rural counterparts. It is noteworthy that these areas are described in development literature as backward, poor and primitive, while in tourist literature the very same regions are presented as idyllic, peaceful and beautiful. Millions of wealthy city-dwellers will spend a substantial proportion of their salary to escape for a few weeks to enjoy life in these 'primitive backwaters'. It is also perfectly normal for the overstressed businessman to seek out precisely the kind of simple village deemed as being 'underdeveloped' as a place to retire. Indeed, it is such a widespread desire that small cottages in rural areas now often cost more than city apartments.

Yet the whole process of industrialization has meant a systematic removal of political and economic power from rural areas and a concomitant loss of self-respect in rural populations. In small communities today, people are often living on the periphery, while power—and even what we call 'culture'—is centralized somewhere else.

In order to see what communities can be like if people retain real self-respect and economic power at the local level, we have to look beyond mainstream accounts. Though such information is not widely publicized, there are numerous books and documents that show what life in largely self-reliant communities was like. I, myself, have written in my book *Ancient Futures*, about the isolated region of Ladakh or 'Little Tibet'—one place that can provide some clues about life in largely self-reliant communities. Unaffected by colonialism or, until recently, development, Ladakh's community-based economy provided people with a sense of self-esteem and control over their own lives. But since the early 1970s, in less than a generation, this culture has been dramatically affected by economic development. Development has effectively dismantled the local economy; it has shifted decision-making power away from the household and village to bureaucracies in distant urban centers; it has changed the education of children away from a focus on local resources and needs towards a lifestyle completely unrelated to Ladakh; and it has implicitly informed them that urban life is glamorous, exciting and easy, and that the life of a farmer is backward and dull. Because of these changes, there has been a loss of self-esteem, an increase in pettiness and small-minded gossip, and unprecedented levels of divisiveness and friction. If these trends continue, future impressions of village life in Ladakh may soon differ little from unfavorable Western stereotypes of small town life.

Urban Sprawl or Revitalized Villages?

An equally common myth that will be cited as an argument against localization is that 'there are too many people to go back to the land'. Interestingly enough, a similar skepticism does not accompany the notion of *urbanizing* the world's population. What is too easily forgotten is that the majority of the world's people today, mostly in the Third World, *are* currently on the land. Ignoring them or speaking as if people are urbanized as part of the human condition is a very dangerous misconception, one that is helping to fuel the whole process of urbanization. That it is considered 'utopian' to suggest a ruralization of the populations of America or Europe, while China's plans to move 440 million people *off* the land and into cities in the next few decades hardly elicits surprise. This 'modernization' of China's economy is part of the same process that has led to unmanageable urban explosions all over the South—from Bangkok and Mexico City to Bombay, Jakarta and Lagos. In these cities, unemployment is rampant, millions are homeless or live in slums, and the social fabric is unraveling.

Even in the North, unhealthy urbanization continues. Rural communities are being steadily dismantled, their populations pushed into spreading sub-urbanized megalopolises, where the vast majority of available jobs are located. In the United States, where only 2% still live on the land, there are now fewer farmers than there are people incarcerated; yet, farms continue to disappear rapidly (Vidal 2000). It is impossible to offer that model to the rest of the world, where the majority of people earn their living as farmers. But where are people saying, "We are too many to move to the city?"

That question is rarely asked because it is implicitly assumed that centralization is somehow more efficient, that urbanized populations use fewer resources. When we take a close look at the real costs of urbanization in the global economy, however, we can see how the opposite is true. Urban centers around the world are extremely resource intensive. The large-scale, centralized systems they require are almost without exception more stressful to the environment than small-scale, diversified, locally adapted production. Food and water, building materials and energy must all be transported great distances via vast energy-consuming infrastructures; their concentrated wastes must be hauled away in trucks and barges, or incinerated at great cost to the environment. In their identical glass and steel towers with windows that never open, even air to breathe must be provided by fans, pumps and non-renewable energy. From the most afflu-ent sections of Paris to the slums of Calcutta, urban populations depend on increasing amounts of packaging and transport for their food, so that every pound of food consumed is accompanied by a dramatic increase in petro-leum consumption as well as significant amounts of pollution and waste.

Precisely because there *are* so many people, a globalized economic model that can effectively feed, house and clothe only a small minority has to be abandoned. It is essential to support instead knowledge systems and economic models that are based on an intimate understanding of diverse regions and their unique climates, soils, and resources.

In the North, where we have for the most part been separated from the land and from each other, we have large steps to take. But even in regions that are highly urbanized, it is possible to nurture a connection to place. By reweaving the fabric of smaller communities within large cities and by re-directing their economic activities toward the natural resources around them, cities can regain their regional character, become more 'livable', and less burdensome to the environment. Our task will be made easier if we support our remaining rural communities and small farmers. They are the key to rebuilding a healthy agricultural base for stronger, more diversified economies.

Resistance and Renewal

A final objection to shifting course is that there is already too much momentum towards globalization, with policy-makers the world over wedded to it. But the scope and potential for public pressure to bring about changes in government policy is actually quite significant, as recent history shows.

A very visible example has been the massive public resistance in Europe against the genetic modification of foods. Despite the attempt of biotech multinationals and the United States government to force GM foods down the throats of European consumers, public pressure to severely restrict or even ban imports of these foods has escalated. As a consequence, it has become impossible for European governments to ignore their voters. In the name of sovereignty and consumers' rights, some of these governments even seem willing to risk a trade war with the US. The four major supermarket chains in Britain, and several others on the continent, have publicly stated that they will not allow genetically modified ingredients to be used in their own brands—again, as a result of enough consumers making their opinions known.

Another, less publicized, victory for citizens has been the stalling of the Multilateral Agreement on Investment. The MAI was an international agreement, written mainly by representatives of transnational banks, global corporations and government trade officials, which aimed to force governments to relinquish much of their power, especially their ability to protect their citizens and maintain social, environmental and health standards. A relatively small number of activists and informed citizens put pressure on governments around the world and forced the stalling of the agreement—a feat made even more impressive by the fact that these negotiations were conducted in total secrecy. Most elected officials, including many ministers, were not even aware of the MAI's existence!

A third example is the US Department of Agriculture's (USDA) retreat from its attempt in 1998 to weaken organic standards, in order to allow large agribusinesses to take advantage of the increasingly lucrative market for organically grown foods. Among other flaws, the proposed rules would have allowed organic foods to be grown from genetically modified seed, fertilized with chemically tainted municipal waste, and sterilized by irradiation—techniques considered 'acceptable' within the global food system, but consistently repudiated by organic farmers. After USDA offices were flooded with thousands of irate letters, calls, and emails from consumers and farmers, the department backed down.

Another encouraging and significant expression of resistance was the mass protest in Seattle during the WTO meeting at the end of 1999. The demonstrations there involved an extraordinary array of farmers, business people, mothers with young children, environmentalists, indigenous people and members of labor unions. Protesters numbered in the tens of thousands, and brought worldwide attention to a process that has over the years taken place in secrecy. The message of the people marching in the streets was very clear: Globalization is not a natural or evolutionary process; it is about specific trade agreements and government policies, and these must be changed. The atmosphere of resistance created by these protests undoubtedly played a major role in the collapse of those talks, and they ensured that future trade decisions—which so fundamentally affect the well-being of the planet and its citizens—are no longer made outside the glare of public scrutiny.

These examples show that ordinary citizens can force changes in policy. Even a relatively small group of well-organized and informed people can make a huge impact. Getting governments to shift course is not impossible, or even unlikely, once enough people understand how disastrous our present course really is.

While the increasingly globalized economy is leading to active resistance on many fronts, it is indirectly spawning spontaneous efforts to reweave the social and economic fabric in ways that mesh with the needs of nature—both ecological and human. Evidence of such changes are emerging everywhere: increasing numbers of doctors and patients are rejecting the commercialized and mechanistic medical mainstream in favor of more preventive and holistic approaches; many architects are finding inspiration in vernacular building styles and are employing more natural materials in their work; awareness of the harmful health and environmental effects of large-scale industrial agriculture is on the rise, and thousands of farmers are switching to organic practices; dietary preferences among consumers are shifting away from processed foods with artificial colorings, flavorings, and preservatives towards fresher foods in their natural state.

It is clear that globalization, with its bias towards concentrating economic activity in fewer and fewer places and businesses, cannot be sustainable. Sustainable Development will only be possible when we resist the forces and policies that are destroying our livelihoods, our cultures, and our environment and make the shift towards localization. We need countless more small, diverse, and local initiatives to support our local economies and communities, which, if supported by policy changes will, over time, inevitably foster a return to long-term sustainability.

Endnotes

[1] www.fas.usda.gov/info/factsheets/mapfact.html
[2] www.corporations.org/system/top100
[3] Based on 150 country reports. See FAO 1996.
[4] Quoted in Bodley J (1982), pp 111–112, *Victims of Progress*.
[5] "Where the Admen Are." *Newsweek*, 14 March 1994, p. 34

References

Bodley J (1982), *Victims of Progress*. Mayfield Publishing, Mountain View CA
FAO (1996) *State of the World's Plant Genetic Resources*. FAO, Rome
FAO (2002) *The State of Food Insecurity in the World, 2002*. FAO, Rome
IMF Statistics Department (1999) *International Financial Statistics, February 1999*, Vol. 53, No. 2. International Monetary Fund, Washington DC
Johnson D (1999) As Agriculture Struggles, Iowa Psychologist Helps his Fellow Farmers Cope. *The New York Times*, 30 May 1999, National Report Pages, p. 12
Lietaer B (1997) Beyond Greed and Scarcity. *Yes! Magazine*, Spring 1997, p. 37
Morris D (1997) When Money Usurps Economy, Something Is Seriously Wrong. *St. Paul Pioneer Press*, 21 October 1997
Murphy S (2002) *Managing the Invisible Hand: Markets, Farmers and International Trade*. Institute for Agriculture and Food Policy, Minneapolis MN. www.iatp.org
Vidal J (2000) eco soundings. *The Guardian* (London and Manchester), 6 September 2000, "Society" section, p. 8
Weiner T (1999) Aid to Farmers Puts Parties in 'Political Bidding Contest.' *The New York Times*, 4 August 1999, p. A14
Wilson EO (1992) *The Diversity of Life*. WW Norton & Co, New York

Business and Human Rights

Klaus M. Leisinger

"Companies cannot and should not be the moral arbiters of the world. They cannot usurp the role of governments, nor solve all the social problems they confront. But their influence on the global economy is growing and their presence increasingly affects the societies in which they operate. With this reality comes the need to recognize that their ability to continue to provide goods and services and create financial wealth—in which the private sector has proved uniquely successful—will depend on their acceptability to an international society which increasingly regards protection of human rights as a condition of the corporate license to operate."[1]

Human Rights: A Business Duty?

Largely unnoticed by the management of most companies, an intense debate has developed in recent years on the issue of 'business and human rights'. But actually the issue is not new, because specialist human rights groups even back in the 1980s linked multinational companies in the extractive sector (oil, diamonds, gold, precious metals) with human rights abuses at their local mining sites.[2] The *Human Development Report 2000* pointed out that 'global corporations' have an "enormous impact on human rights—in their employment practices, in their environmental impact, in their support for corrupt regimes or their advocacy for policy changes" and called for "corporate human rights standards, implementation measures, and independent audits." (UNDP 2000) The *OECD Guidelines for Multinational Enterprises*, revised in 2000, already included a reference to human rights (OECD 2000). What is new, however, is the dynamic increase over the last three years or so in the breadth and depth of the general business-related human rights debate.[3]

This may be unexpected, but it does not pose a problem for companies competing with integrity (see de George 1993). Today, all actors of civil

M. Keiner (ed.), The Future of Sustainability, 117–151.
© 2006 *Springer. Printed in the Netherlands.*

society must perceive a responsibility for human development and thus respect the equal and inalienable rights of all members of the human family
as enshrined in the *Universal Declaration of Human Rights* (UDHR)
adopted by the General Assembly of the United Nations on 10 December
1948. 'Good' corporations and those responsible for their corporate conduct will therefore perceive a duty to support and respect human rights and
do their utmost to ensure that the spirit of the *Universal Declaration* is upheld in their sphere of activity and influence. At a minimum, they will refrain from actions that obstruct the realization of those rights. Where, then,
are the problems?

The present debate suffers from the fact that, at one end of the political
opinion spectrum human rights activists create the impression that the
whole misery of people in developing countries is largely the consequence
of cynical orgies of human rights abuses by multinational companies (e.g.,
CETIM and American Association of Jurists 2002). At the other end of the
spectrum, institutions with close ties to business go on record saying that
there are no business-specific human rights problems, as all demands of
the *Universal Declaration of Human Rights* are directed exclusively at the
state and its regulatory authorities (International Chamber of Commerce
and International Organization of Employers 2003). Human rights demands being made of other actors in society (e.g., of business enterprises),
they say, detract from the actual perpetrators—widely known despots and
their entourage abusing basic human rights within their sphere of power.

The tenor of the human rights debate has become increasingly critical of
'transnational corporations' due to a deep-seated disquiet about globalization. Opinion polls show that nine out of ten respondents who are interested in development policy and work in NGOs or who have close ties to
that work see too much globalization emphasis on trade and investment
and far too little attention being paid to human rights or other non-
economic issues.[4]

Against this background, the *UN Global Compact* (UNGC) initiative of
UN Secretary-General Kofi Annan continues to be of major importance.[5] It
takes up this disquiet and aims to counteract it by encouraging companies
to commit themselves publicly to compliance with certain minimum standards of a political, social, and ecological nature. Convinced that weaving
universal values into the fabric of global markets and corporate practices
will help advance broad societal goals while securing open markets, Annan
challenged world business leaders to 'embrace and enact' the *Global
Compact,* both in their individual corporate practices and by supporting
appropriate public policies and promoting fair business practices. Such
fairness is expressed by good labor standards and enlightened protection of
the environment—but also by corporate efforts to 'support and respect the

protection of the international human rights within their sphere of influence' as well as 'make sure their own corporations are not complicit in human rights abuses'.

The principles relating to social and ecological issues were not a problem for companies following state-of-the-art practices of 'good corporate citizenship' or 'corporate social responsibility'. The two human rights principles, however, led them into territory that was new and unfamiliar to the management of most companies.

At about the same time, a Sub-Commission of the Human Rights Commission (2003) started its work to develop a set of *UN Norms on the Responsibilities of Transnational Corporations and other Businesses on Human Rights.*[6] One of the central aims of this Sub-Commission was to strengthen and put into operation the two human rights principles of the UNGC. While it is true to say that the UN norms represent a welcome strengthening of the two UNGC principles in terms of content, they are unclear—at least in their present form—on a number of important procedural issues.[7]

Whatever the tune of the debate is, the fact remains that business enterprises have a moral obligation to respect human rights. If they do not comply with these most essential elements of their social responsibility, they surely risk their societal (if not legal) license to operate. Hence, at least the management of enlightened companies see themselves confronted with the question of how to respond to the increasing importance of human rights demands on business enterprises.

What Constitutes a Fair Societal Distribution of Labor?

Modern societies are highly complex systems of human coexistence. They contain a multitude of actors (individuals, groups, organizations) whose interests, objectives and differing modes of behavior and regulations are entwined in a circularly interdependent relationship.[8] To meet the superordinate aims of a modern society (such as respect for and fulfillment of human rights), the various actors must contribute according to their resources and abilities in the context of a social contract (see, e.g., Donaldson and Dunfee 1999).

With many demands made on business enterprises by interest groups, it is not the question of their fundamental justification that is at issue. The issue that is contentious is the question of who is the appropriate bearer of duty. To demand rights without determining at least the direction for collateral duties creates the risk of raising expectations that cannot be fulfilled. No societal actor has all the obligations to bear and no one enjoys all the rights. This is especially true in the context of human rights.

Paul Streeten (2003) helps to sort out the different duties by distinguishing four different areas of human rights:

– In the narrowest sense they are *rights of personal integrity*: the right not to be tortured or killed, not to be imprisoned without due process of law. Freedom of conscience and expression, freedom from arbitrary deprivation of liberty, the right to assembly, and freedom of association belong to them. These rights apply under all governments, irrespective of their political color—and take little more resources than the political will.

– A second group consists of *civil rights*, or what in Anglo-Saxon countries is described as the 'rule of law' and in Germany as the 'Rechtsstaat'. This group comprises the rights of citizens against their government. The rulers themselves are subject to the law.

– In the third group are *political rights*. These enable citizens to participate in government by voting for their representatives, throwing them out, and restricting those elected in what they can do while they are in power. Representation can take many forms, of which one adult person/one vote or a multiparty system is only one.

– The fourth group is that of *economic, social, and cultural rights*, embodied in the *Universal Declaration of Human Rights* of 1948 and the *International Covenant on Economic, Social, and Cultural Rights* of 1966. Economic, social, and cultural rights are positive rights to scarce resources and therefore distinct from the negative rights not to have certain things done to one. In the case of cultural rights, conflicts can arise between the rights of communities and of individuals. The rights to universal primary education, to adequate health standards, to employment, and to minimum wages and collective bargaining are quite different from negative rights. It has taken the more enlightened advanced societies three centuries to achieve the civil, political, and social dimensions of human development.

The interpretation of economic and social rights can therefore not be made irrespective of the stage of development of a country, its available capabilities and resources, and competing claims on these resources. Social services (such as support for unemployed citizens through dole systems) that have been financed by a mature European nation such as Germany in the past 20 years seem no longer sustainable there—and they are well out of reach for the finance capability of any sub-Saharan African nation. It is therefore recommended to distinguish in this context, as Streeten (2003) has done, between

> *"... aspirations, which are ideals we hope to attain eventually, and rights, about which there is something absolute ... by calling some human aspirations a right, the objective in question has been given a moral absolute and categorical supremacy, irrespective of the nature of the right, its appropriateness to the circumstances in which it is proclaimed, or to the possibilities or costs achieving it."* (Streeten 2003)

There is no doubt that all actors in every civilized society should aspire to opportunities for comprehensive human development, including such rights as expressed by Article 24 (such as the right to rest and leisure), Article 25 (right to a standard of living adequate for health and well-being of the individual and the family, including food, clothing, housing, medical care, and necessary social services, and the right to security in the event of unemployment, sickness, disability, widowhood, old age, or other lack of livelihood in circumstances beyond a person's control), and Article 26 (right to education) of the UDHR. It would be unrealistic, however, to expect poor nations to guarantee full and immediate implementation.

Accepting this poses highly political questions: If those carrying the primary responsibility to deliver on these rights are not able to do so, who is next in line of responsibility? One thing is sure: A flourishing economy that benefits all social groups—carried by flourishing enterprises—is the most important prerequisite for the satisfaction of basic needs and the fulfillment of economic, social, and cultural rights. In terms of responsibility for pursuing pro-poor economic development, restructuring budgets to provide adequate expenditure for primary human concerns, ensuring participation and social reforms, protecting environmental resources as well as the social capital of poor communities, and securing human rights in law, there is clearly one duty bearer: the nation state.

Characteristics of Modern Society

According to Niklas Luhmann, a characteristic feature of modern society is its differentiation into a variety of functionally specialized subsystems, such as economy, law, politics, religion, science, education, etc. None of these subsystems of society are able to substitute one another, let alone all others: "Function systems cannot step in for, replace or even simply relieve one another." (Luhmann 1989) The quality of cooperation of the different subsystems determines the degree of possible synergies and allows for the whole (society) being more than the sum of its parts (subsystems).

The case of the economy illustrates the limitations of a subsystem. Money is the medium of communication in the modern economy. For Luhmann the economy is the totality of those operations that are executed by payments. Business enterprises and individuals engaged in business form the economic subsystem. Its prime function is to ensure that the needs of all citizens are satisfied as cost-effectively as possible. The self-interest of individuals and competition are the organizing forces in this process (von Nell-Breuning 1990). Decisions are made according to the economic efficiency principle (benefit-cost ratio) and other criteria of business rationality. This results in rules of conduct that differ fundamentally from those of other societal subsystems. As a function subsystem of modern society the economy can treat problems only insofar as they are communicated as matters of economic costs and benefits, otherwise it will regard them as 'noise' and refrain from such operations (Luhmann 1995). To apply rules and values of other subsystems is not suitable for the functional capacity and effectiveness of the economic sub-system. Redistribution policy or social transfers of results in the name of charity, for example, belong to the functions of other societal subsystems. The economic subsystem acts on and through markets; a compassionate anti-economy could not be sustained owing to constraints inherent in the system. Under 'normal' circumstances, i.e., a functioning society and good governance, other subsystems assume responsibility for the issues that cannot be dealt with by the economic subsystem.

It is a subject of debate in modern societies as to what a self-evident duty is for a company and what constitutes an unreasonable demand. Different stakeholders define the responsibilities of business enterprises differently as a result of their differing values and interests. Not every demand of every stakeholder becomes the moral duty of the company. In defining the corporate social responsibility of companies—the human rights dimension is just one of several—a distinction should be drawn between three dimensions of responsibility involving differing degrees of obligation, namely:

– The '*must*' dimension—non-negotiable essentials incumbent on the respective industry that, by general social consensus, go without saying;

– The '*ought to*' dimension—describing good corporate citizenship standards through the application of internal guidelines for sensitive business areas, particularly in countries where the quality of the law is insufficient or law is not enforced; and

– '*Can*' rules – the assumption of responsibility in an even further dimension.

The 'Must' Dimension of Societal Responsibility

The dimension of responsibility that is absolutely essential for companies is not much different today than in the past. A company has to produce good-quality goods and services for which there is a demand from potential buyers with purchasing power, and to sell them at profitable prices. In the process, the business enterprise—like all other societal actors—has to comply with all the laws and regulatory requirements as well as respect relevant customs. To conduct its business activities, the enterprise creates and maintains healthy and safe jobs, pays employees competitive wages, and treats them fairly. The 'must' dimension of corporate activity also includes securing a fair interest on the capital invested by the owners of the company: the shareholders.

Other essential responsibilities include the protection of the environment, contributions to pension funds and insurance systems, and paying taxes. With the taxes paid, the state should finance its operations and fulfill its tasks. Since most companies also provide training and education, further value-added accrues to society. Added value also arises from the productive use of products and services. In the case of a pharmaceutical company, considerable benefit to society arises from the use of medicines because mortality is reduced and sickness and disability are cured or relieved.

By accepting this essential dimension of corporate social responsibility as a matter of course, a company contributes to the fulfillment of economic, social, and cultural human rights of citizens just by doing business under 'normal' circumstances. By 'normal' circumstances I mean that the countries in which corporations operate are characterized by good governance, i.e., legitimized exercise of political power; correct financial, economic, social, and other policy decisions; the rule of law; the rational allocation of resources, etc. Unfortunately, circumstances in many countries are not 'normal'. The most difficult human rights problems occur in countries in which the state and its organs are either not able or are unwilling to meet their responsibilities. Under such circumstances, a company is well advised to go beyond the 'must' dimension of human rights and other societal responsibilities.

The 'Ought to' Dimension of Societal Responsibility

The Green Paper *Promoting a European Framework for Corporate Social Responsibility* defines the 'ought to' dimension as the actual social responsibility of companies and as a

"... concept whereby companies integrate social and environmental concerns in their business operations and in their interaction with their stakeholders on a voluntary basis." (European Commission 2002)

If we understand the 'must' dimension as compliance with legal and conventional modes of behavior, then the 'ought to' dimension can be seen as the constructive and generous filling of unregulated space as proposed, for example, in the *UN Global Compact*.

Enlightened labor and environmental standards are applied even if local law and regulations would allow for lower standards. The reason for adherence to the enlightened spirit of the rule and not just the letter of actual national law is based on values: *'It is the right thing to do'*. By adopting corporate citizenship guidelines, enlightened companies create a framework of self-commitment, which guarantees legitimate business behavior even if the local legal preconditions are lacking. Since these activities are voluntary and therefore to a certain degree dependent on the financial muscle of a company, the 'ought to' dimension of social corporate responsibility is fulfilled in different ways by different companies and over time. This is even more so in the case of the 'can' dimension.

The 'Can' Dimension of Societal Responsibility

The 'can' dimension of social responsibility describes special activities that are neither set out by law nor customary in the industry—and yet that can be of substantial benefit to people. A company, for example, may offer free or subsidized meals to its employees, free or subsidized transport, free or subsidized kindergarten facilities for children of working mothers, or free further training opportunities using the company's infrastructure, or scholarship programs for the children of employees in lower-income groups. Special activities for diagnosis, therapy, and psychosocial care, such as for employees suffering from HIV/AIDS, also fall into this category.

The establishment and funding of foundations with a philanthropic mission also come into the 'can do' category. In addition to financial resources, some companies have knowledge and experience that they can deploy in projects and programs of development cooperation and humanitarian aid. These programs would greatly benefit from companies in terms of effectiveness, efficiency, and significance.[9] The creative fantasy in the context of the 'can' dimension knows no boundaries.

The Human Rights Dimension of Corporate Social Responsibility

Several companies can be lauded for their efforts to integrate an ethical approach to multinational operations. While none have been specific in stating that they have used the *Universal Declaration of Human Rights* developed by the United Nations[10], it is not difficult to see that its incorporation is possible, because they accept a number of human-rights-related duties.

Obligations in the Context of Civil and Political Human Rights

Although the civil and political rights of the UDHR are above all incumbent on states and their institutions, the preamble of the UDHR stipulates that 'every individual and every organ of society' is called on to respect and promote these rights. Business enterprises are 'organs of society' and enlightened corporations therefore accept—to different degrees—human-rights-related responsibilities. The prime responsibility is to ensure that a company's activities do not contribute directly or indirectly to civil and political human rights abuses and that the company under no circumstances will knowingly benefit from such abuses. This implies that a corporation informs itself of the human rights impact of its principal activities and major strategic decisions so that it can avoid complicity in human rights abuses.

Obligations in the Context of Economic, Social, and Cultural Human Rights

The respect and promotion of economic, social, and cultural rights is also primarily a duty of the state and its regulatory authorities. Whereas civil and political rights are defensive rights that aim to prevent state interference with individual freedoms, economic, social and cultural rights are positive rights and are more difficult to enforce, as their implementation requires the material support of duty bearers.

Today, far from all rights based upon the Universal Declaration of Human Rights and set out in the International Covenant on Economic, Social, and Cultural Rights are fulfilled for a large number of poor people. The lives of more than 1.2 billion people living in absolute poverty are characterized by the sad fact that their right to adequate food, clothing, and housing as well as their right to the highest attainable standard of physical and mental health remain unfulfilled.[11] The sheer dimension of today's global poverty problems makes it obvious that private companies can only contribute toward the support and respect of the economic, social, and cultural rights in the context of their normal business activities (see also International Council on Human Rights Policy 2003). Economic and social rights such as the right to work (Article 23 of the UDHR), the right to a standard of living adequate for the health and well-being of a human being and his or her family, including a right to medical care (Article 25), and the right to education (Article 26) cannot be progressively implemented without good governance, effective public services, and appropriate allocation of resources. Corporate contributions toward the fulfillment of such rights come through by employing people, paying fair wages, and providing social benefits as well as by creating economic value-added through their normal business activity while obeying all laws and regulations. Through specific human-rights-related corporate citizenship guidelines and their incorporation into normal business activities, companies contribute to the fulfillment of various economic, social, and cultural rights.

Right to Equal Opportunity and Non-Discriminatory Treatment

Enlightened corporations have non-discriminatory business policies, including but not limited to those relating to recruitment, hiring, discharge, pay, promotion, and training.[12] All employees are treated with equality, respect, and dignity. Discrimination based on race, color, sex, religion, political opinion, nationality, social origin, social status, indigenous status, disability, age (except for children, who may be given greater protection), or other status of the individual unrelated to the individual's ability to perform a job are not tolerated within the sphere of the corporation's influence. Nor is intimidation or degrading treatment tolerated. No employee is disciplined without fair procedures.

Right to Security of Persons

Responsible corporations will not engage in or benefit from war crimes, crimes against humanity, genocide, torture, forced disappearance, hostage-taking, other violations of humanitarian law, and other international crimes against the human person as defined by international law.[13] Their security arrangements observe international human rights norms as well as the laws and professional standards of the country in which they operate and are used only for preventive or defensive services. Security personnel are instructed to only use force when strictly necessary and only to an extent proportional to the threat.

Rights of Workers

Responsible corporations do not use forced or compulsory labor and respect the rights of children.[14] Workers are recruited, paid, and provided with working conditions that meet or exceed the package necessary to cover basic living needs ('living wages'). All workers and employees have the right to choose whether to join a trade union or employee association. Responsible corporations also make special efforts to respect the rights of children to be protected from economic exploitation[15]—they do not employ any person under the age of 18 in any type of work that by its nature or circumstances is hazardous, interferes with the child's education, or is carried out in a way likely to jeopardize the health, safety, or morals of young persons.

For all employees and workers, a safe and healthy working environment is provided at least in accordance with the national requirements of the countries in which they are located and with international standards.[16]

Respect for National Sovereignty and Local Communities

It goes without saying that corporations competing with integrity will recognize and respect all norms of relevant international and national laws and regulations as well as the authority of the countries in which its group companies operate. Within the limits of its resources and capabilities, such companies strive to encourage social progress and development by expanding economic opportunities. Enlightened corporations respect the rights of local communities affected by its activities and the rights of indigenous peoples and communities consistent with international human rights standards.

Obligations with Regard to Environmental Protection

In many respects, socioeconomic development facilitates the enhancement of human capabilities, which in turn help to secure basic freedoms and realize human rights. The use of nature may on the one hand be the price for economic development and on the other hand may make economic growth unsustainable.[17]

As the environmental impact of business activities differs substantially between sectors (crude oil production versus insurance enterprises, for example), every sector has specific environmental problems to solve and obligations to bear. It is, however, inconceivable that responsible corporations would not carry out their activities at least in accordance with national laws, regulations, administrative practices, and policies relating to the preservation of the environment of the countries in which they operate. As with social standards, enlightened corporations voluntarily set higher standards if the standards required by local law and regulation do not meet the corporation's understanding of environmental stewardship.

Other Obligations

In accordance with the state of the art of business ethics, enlightened corporations do not offer, promise, give, accept, condone, knowingly benefit from, or demand a bribe or other improper advantage. Long-term self-interest demands corporate actions in accordance with fair business, marketing, and advertising practices and takes all necessary steps to ensure the safety and quality of the goods and services they provide. Such corporations therefore adhere to the relevant international standards of business practices regarding competition and anti-trust, and ensure that all marketing claims are independently verifiable, satisfy reasonable and relevant legal levels of truthfulness, and are not misleading.

Entrepreneurial Options

Companies respond in different ways to political challenges, depending on corporate culture, historical experiences, or the philosophy of top management. In the context of the human rights debate, the management of a company has in principle three options for action:

- Defend the perceived status quo or even actively resist

- Duck, wait, and hope for the best

– See the human rights debate as an opportunity for corporate citizenship leadership

Defend the Perceived Status Quo: Human Rights are not 'The Business of Business'

There are a number of understandable reasons why managers of companies would not feel that demands of a human rights nature on their company have anything to do with them. Firstly, they usually associate 'human rights' with civil and political human rights only (Articles 1–21) and not with economic, social, and cultural human rights (Articles 22–29) of the *Universal Declaration of Human Rights*. Secondly, they see the state and its institutions as the primary bearers of duty, and not private companies. Thirdly, even managers who are empathetic to human rights concerns are surprised that at a time when the most horrific abuses of the most fundamental human rights—the right to life and to freedom from bodily harm—are documented almost daily from notorious countries,[18] it is the business community that is the focus of interest with regard to human rights policy. Last but not least: A considerable portion of the debate on 'human rights and business' consists of completely non-discriminating and generalizing charges against businesses, and this makes many concerned people from the private sector hesitant to engage in a debate that could turn out to be too politicized to yield constructive results.[19]

Indeed, it is primarily the implementation of national and international law through responsible government that is called for as a way out of existing human rights deficits. Without the acceptance of essential national obligations, no sustainable and essential progress can be achieved for those people whose shattering destiny we are familiar with from the annual reports of Amnesty International. But one thing does not preclude the other: The commitment of the state and its bodies does not exclude the assumption of responsibility by 'other organs of society'.[20] On the contrary, this becomes a duty precisely when the holders of state power are not able or willing to protect the citizens of a country from violation of their rights. Looking at the annual reports of Amnesty International, it is precisely in places where the state fails to meet its primary responsibilities that the potential vulnerability for companies is particularly high, because they have to operate in an extremely difficult sociopolitical environment (Davis and Nelson 2003):

– Inadequate legal frameworks and governance structures to ensure fair and equitable administration of justice and regulations

- Weak, authoritarian, or failing public sector institutions with thriving corruption

- High levels of poverty and inequality in the distribution of resources and livelihood opportunities

- Lack of access to basic services such as education, health care, energy, water and sanitation, and telecommunications

- Strict press controls

- Existing or potential civil conflict with politically or ethnically motivated human rights violations

Lack of good governance gives rise to a vacuum that has to be responsibly filled by other actors in society, including companies: only in this way can they minimize the risk of not becoming part of the problem themselves. But more on this later.

Duck, Wait, and Hope for the Best

Most multinational corporations have so far not responded to the human rights debate, at least not visibly or audibly.[21] It seems as if they are waiting until the 'discussion caravan' has moved on and the globalization debate has turned its attention to other issues. If—unexpectedly—the debate is ultimately to have consequences for national legislation, then a company just has to do what cannot be avoided (any longer) because of the new laws. Other arguments brought forward in favor of ducking and waiting relate to a fear similar to James Duesenberry's 'ratchet effect': Corporate performance within the framework of a corporate citizenship policy—that is, beyond what is stipulated by law—could, by virtue of the normative force of what has become fact, establish a performance level below which it is no longer acceptable to fall and that becomes the baseline for additional (including legal) demands.[22]

In my perception, companies that opt for the duck-and-wait strategy unfortunately have a good chance of success. In many cases, the attention of the critical public is focused not so much on those companies that use local deficits of the law and refuse to engage in debate but on those that behave responsibly, face up to the company-related human rights debate, and take an active part in it with arguments of their own. The more prominent a company, so it seems, the greater the size of the sounding board for critics, more or less regardless of the severity of the issue that is being criticized. This being so supports the rationale that the crucial element for a management deciding to deliver corporate performance standards beyond the legal minima must be its conviction that it is 'the right thing to do'.

The situation is different for companies that have signed up to the UN Global Compact—in doing so, they have already committed to two human rights principles. Since these are relatively open in their wording, such companies only have the choice of defining for themselves what commitments they believe they have entered into—or of leaving this interpretation to other actors of civil society and then finding themselves confronted with demands that they either have to or want to reject. As this would be a strategically poor choice, the right thing to do is to take an active part in the human rights and business debate and to perceive this as an opportunity for leadership in corporate citizenship practices.

See the Human Rights Debate as Opportunity for Corporate Citizenship Leadership

Successful business enterprises are organized for constant change and innovation—and the current focus on human rights is just one of those changes. As Peter Drucker observed many years ago, successful companies are those that focus on responsibility rather than power, on long-term success and societal reputation rather than piling short-term results on top of short-term results (Drucker 1993).[23] Getting the human rights dimension right—right in the sense of an enlightened balance between the common good and enlightened corporate self-interest—is no longer only a question of moral choice but increasingly an important asset on the reputation market created by a growing part of global civil society. Good managers realize that it will be very difficult to be a world-class company with a second-class human rights record—and they act upon this (Avery 2000).

Companies that signed up to the principles of the UN Global Compact will set in motion an internal process of definition and implementation with regard to all commitments entered into, the human rights obligations being one of them (Leisinger 2003). A process of this kind has different phases: reflection and consultation, discussion and decision-making, and implementation.

Reflect and Consult

Rory Sullivan describes the intersection between human rights and business as

*"Chaotic and contested: on the one hand there are those who see com-
panies as 'the source of all evil'. On the other are those who have a
touching faith in the abilities of companies, economic growth and 'the
market' to resolve all of these human rights problems. Yet the truth,
if there is such a thing, is far more complex and indeterminate than either
of these extreme perspectives allows. Despite the increasing use of human
rights language in public policy discourses, the expectations of companies
remain unclear."* (Sullivan 2003)

Indeed, few companies had any clear idea before signing up to the *UN
Global Compact* what it meant for them 'to support and respect the protection of international human rights within their sphere of influence' as well
as 'to make sure their own corporations are not complicit in human rights
abuses'. In normal cases—that is, in the case of companies operating
within the law and committed to basic ethical values—there seems to be an
intuitive assumption there are no human rights violations through their
own activities and therefore no problems were to be expected. This assumption largely corresponds to my experience. Much of the action that is
demanded today in the human rights debate already forms part of the social and ecological management processes of enlightened companies.

It is nevertheless inadvisable to carry on as if nothing had happened
without any further reflection; a deeper consideration of the problems is
called for. On the one hand, the issues around the 'human rights and business' debate are defined substantially broader by many stakeholders than
most managers assume. On the other hand, as far as actual human rights
performance is concerned, it is not wise to operate with assumptions where
empirical knowledge can be gained. The more facts that can be ascertained
on sensitive issues and the more insight there is on the existing pluralism
of values with regard to the facts, the better the decision-making basis for
informed policy choices. What need to be answered first and foremost are
questions such as 'What could potentially sensitive aspects of the business
activity be?', 'Where do stakeholders outside the company see potential or
actual issues of relevance to human rights in the context of our business
activity?', and 'Are there vulnerabilities that arise through cooperation
with others and, if so, how do we cope with this?' A conscious human
rights assessment if not human rights audit of current corporate practices
might be recommended if a rough assessment does reveal unexpected
negative surprises.

The intense search for answers to these complex questions triggers important sensitizing effects within the company, especially for managers
whose area of responsibility is confined in the day-to-day routine of work
to purely business or financial functions or, depending on the field of

work, to biological, chemical, or other matters. The very fact that human rights issues are discussed internally on the company's own initiative—not out of a defensive compulsion—and that critical questions are posed increases the corporate social sensibility and hence competence.

In internal and external consultation processes, the broadest possible spectrum of opinion needs to be obtained. Managers whose workplace is located in countries with a poor human rights performance can make hugely important contributions to the discussion that are of relevance to day-to-day practice. Procurement managers have a view of things that differs from that of communications officers, and the legal department in such a process has still different functions:

"General Counsels are paid to worry about possible threats of litigation, however remote." (Schrage 2003)

The legal view of things is becoming increasingly important if only by virtue of the way in which US courts currently interpret the *Alien Tort Statute* of 1789. A serious analysis of potential vulnerabilities and corresponding guidelines for corporate activities in sensitive areas are a credible first 'good faith effort'. Best practices are always anticipatory—a proactive approach, however, presupposes appropriate reflection on different fact and value scenarios.

Internal consultation processes are also necessary to broaden ownership for what are at least initially 'non-mainstream' positions: anyone who wants to change paradigms of corporate policy must create majorities for the envisaged changes within the company through persuasion. Experience shows that when something is perceived as being imposed 'from above' it will have little effect in daily practice. If a policy change is perceived as a threat (to investment plans, marketing policy, customer relations, and so on), it may—despite the decision being taken at the level of corporate policy—lead to passive resistance, cover-up practices, and refurbishment for some Potemkin façades. All this makes it more difficult to retain a dispassionate grasp of the essentials—and thus to make a rational analysis of the status quo.

Since intra-institutional analyses always involve the risk of being self-referential and therefore of leaving out important aspects from the analysis, external consultations contribute to a better basis for decision-making. This is especially true in the case of complex political judgments such as company-specific human rights issues. Not only is it wise to use the knowledge and experience of specialized non-governmental organizations (NGOs) for a company's own decision-making processes, society's pluralism of interests also gives rise to opportunities. Potentially fatal deficits of perception arise where people or institutions confuse their view of things with the things themselves. Sustainable solutions to complex problems normally transcend the initial preferences of corporate management, taking into account the differing life experiences, value premises, and constellations of interest to improve the quality of the eventual decision. Specialized interest groups are best able to present the relevant portfolio of values, to articulate special interests, and to show ways to preserve them.

Discuss and Decide

After a good decision-making basis has been developed through broad consultation and deep reflection, an intensive internal discussion process must follow. Weighing the pros and cons of different options and wrestling with them to come up with what will be the corporate position to be implemented contributes to a further sensitization within the company. This in turn increases the ability to understand corporate responsibilities and their limits. The fact that differing views are presented not only between outside stakeholders and the management but also within management itself should be seen as an opportunity to improve mutual understanding and as a chance for better solutions. Precisely in the case of politically prestructured questions, it would be dangerous for a company to reach a concluding judgment too early based on the personal preferences of individuals. Here, too, the principle applies that consultation of people with different personal inclinations or professional or cultural backgrounds, different value judgments and experiences of life, or other characteristics that influence their judgment enriches the debate and thus enhances the quality of the decision-making.

The consultation process must also be used to clarify ambiguous terms (such as 'sphere of influence', 'complicity', and 'precautionary principle'). 'Sphere of influence' in the context of the Global Compact is relatively clearly understood as the core operations, the business partners, and the host communities.[24] Mary Robinson refined this concept with the remark

"Clearly, the closer the company's connection to the victims of rights violations, the greater is its duty to protect. Employees, consumers, and the communities in which the company operates would be within a first line of responsibility."[25]

Analyzing the wide variety of possibilities for the definition of 'complicity', it seems much more difficult to come to an accepted corporate understanding of the concept (Clapham and Jerby 2001).[26] And yet differentiations are possible: With a small number of known corporate common sense measures, it should be possible to rule out 'direct complicity' in the sense of consciously assisting a third party in violating human rights. A well-informed and sensitive management should also be able to avoid 'beneficial corporate complicity'—defined as benefiting directly from human rights violations of a third party.

Staying clear of 'silent complicity' is a bigger challenge, as this notion reflects the expectation on companies that they raise a certain quality of human rights violations with the appropriate authorities. To speak out about human rights, whether in corporate management development courses, contract negotiations with third parties or at other occasions helps create a business environment that supports the protection of human rights. Individuals working in corporations may raise human rights issues at private meetings with higher-ranking officials, politicians, or ministers—even there, diplomatic suggestions may achieve better results than overt criticism. Many companies, however, do not encourage their managers to adopt a highly political role while on corporate duty. They see no corporate mandate to act as a vehicle for global diplomacy. As public perceptions of corporate behavior might differ significantly from corporate perception regarding 'silent complicity', a position paper is advisable on this topic.

The 'precautionary principle' is another term that is used widely but differently by stakeholders. The concept was developed in the context of the UN Conference on Environment and Development in 1992 and dealt exclusively with environmental issues around global warming. Today there are a number of vague definitions and demands, such as

"... the burden of proof of harmlessness of a new technology, process, activity, or chemical lies with the proponents, not with the general public." (See Montague 1998)[27]

But what does 'harmlessness' mean for a research-based company working with pharmaceutical compounds active, for instance, against cancer but not at all 'harmless' as far as side effects are concerned? Even if different stakeholders (mis)use the precautionary principle in widely different contexts (such as against the use of genetic engineering for Third World agriculture), a corporate definition of this principle in the original environmental context seems appropriate.

The use of the 'precautionary principle' implies a certain way of thinking which ought to be made transparent if the use of this principle is suggested: There are two points of view when we face risk and uncertainties. One is based on the precautionary principle. The precautionary principle says when there is any risk of a major disaster, no action should be permitted that increases the risk. If, as so often happens, an action promises to bring substantial benefits together with some risk of a major disaster, no balancing of benefits against risks is to be allowed. Any action carrying a risk of a major disaster must be prohibited, regardless of the costs of prohibition.

The opposing point of view (which is mine) holds that risks are unavoidable, that no possible course of action or inaction will eliminate risks, and that a prudent course of action must be based on a balancing of risks against benefits and costs. The precautionary principle asserts that faced with a possibility, however remote, of some catastrophic development, prudent policy demands that whatever action is required to prevent it be taken. This implies that the required action has to be taken however high the costs. When we buy a padlock to prevent our bicycle being stolen, we compare the value of the bicycle with the chances of theft and the costs of the padlock. If the bicycle is worthless and the cost of the padlock is very high, we don't buy it.

To apply the precautionary principle means that irrespective of the chances of future loss, the scale of the loss and the costs of preventing it one must incur these costs. We have the choice between (1) accepting some remote and unquantifiable possibility of severe effects and (2) certain catastrophes if policies are adopted to avoid it. The economic costs of avoiding all conceivable possibilities of a major disaster could be astronomic.[28]

As a result of the discussion processes, the company has a better understanding of all human-rights-related aspects of its activities and is able to decide in the best interest of the matter. Different issues will have a different weight and importance for different sectors (such as oil, textiles, banks, data processing, or pharmaceutical industries). Even within a specific sector (the pharmaceutical industry, for example, and there between research-based and generics companies), different problems will lead to different decisions regarding corporate human rights policy. For companies competing with integrity in all sectors, however, it should be possible to develop a relatively broad basic corridor for their human rights performance.

Implement

Once a company has self-regulated the details of its corporate human rights endeavors, a 'normal' management process has to be implemented—that is, compliance with the human rights guidelines becomes part and parcel of normal business activities. The usual process parameters for this are:

– Appoint a senior manager to be in charge of the human rights responsibility, including mainstreaming and supervising the human rights strategy throughout the corporate world;

– Initiate an interactive communication strategy for all employees (not only the management) and develop an attractive roll-out campaign in different languages to enhance interest in the issues;

– Provide internal training of key personnel worldwide, using case studies of relevance to corporate business and a tool box (including dilemma sharing); involving relevant NGOs to provide an 'out of the box view' adds to the quality of such endeavors;

– Develop 'measurables' in the sense of qualitative and quantitative benchmarks that are relevant to the human rights debate of the sector the company belongs to;

– Set performance targets for sensitive responsibility areas (such as security and human resources) and link achievement to income of responsible managers;

– Ensure compliance monitoring throughout the corporation with special emphasis on potential vulnerabilities of corporations in that sector;

– Develop and implement external verification mechanisms;

– Report on the success as well as the failure of performance as well as other activities according to sector of activity.[29]

As such a complex implementation process will take some time, the management should set milestones to keep track. If this is done, the promise 'to support and respect the protection of international human rights within their sphere of influence' as well as 'to make sure their own corporations are not complicit in human rights abuses' becomes a self-evident part of normal business activity.

Open Issues for Discussion in a Learning Forum

The 'human rights and business' debate has progressed a lot in the past years. There are, however, still a number of open issues that need further debate in good faith among different stakeholders. Four will be touched on here:

- How far does the corporate arm reach?
- What is an appropriate verification process of corporate human rights performance?
- What are useful indicators?
- How can we develop a human rights-related 'Richter Scale'?

How far does the Corporate Arm Reach?

No reasonable person would dispute that corporate activities must be brought out in a manner that upholds the rights of employees and workers as well as of the local communities they are active in. The issue at stake is to define reasonable boundaries on the human rights responsibilities of business enterprises. It is relatively easy to determine where it begins: A company should adopt explicit corporate guidelines on human rights and establish procedures to ensure that all business activities are examined for the human rights content. Wherever a company has direct control (that is, predominantly in its own operations), it can be held accountable for its human rights record. This includes the moral duty to protect the rights of employees against illegitimate interference of local authorities, for example by providing them with legal assistance in cases of them suffering from violations of their civil and political rights.

It gets a bit more difficult to exert indirect control—to influence suppliers, subcontractors, and business partners to adhere to the spirit of someone's corporate responsibility standards, including human rights. But there are ways (processes) and means (carrots and sticks) to correct deficits and initiate policy changes that prevent them in future.

But where are the limits? What is the permissible nature and extent of corporate contributions toward the creation of an enabling environment for the realization of human rights in states whose human rights performance gives raise to justified criticism and where local standards conflict with international norms. In contrast to the long-standing[30] disapproval of transnational corporations interfering in domestic political affairs, recent thinking from advocacy groups and well-intentioned NGOs seems to call for corporate involvement in human-rights-related political activities in 'difficult' countries.

It is relatively easy to determine that at the very minimum, corporations have a moral obligation to ensure they do not undermine elected governments or the democratic process, but it is much more difficult to draw the line about where direct interference in the political process—such as against repression of religious, ethnic, or political opposition groups—is acceptable and where it is inappropriate. Should a multinational corporation working in a country that does not allow or that severely restricts unions contravene local laws? Should General Motors contribute to the fulfillment of Article 18 of the UDHR (the right to freedom of thought, conscience, and religion) by permitting Falun Gong meetings at its Shanghai plant? (See Litvin 2003) Is a 'good' company expected to close down its plants in a country after an undesirable change of government from a legitimate one to a human-rights-abusive regime? Would you have to ask the workers who are direct victims of a boycott decision—mostly poor people, who anyway suffer most from such regimes—whether they accept a deterioration of their personal life conditions as the price of an external pressure that could lead in the long run to an improvement of the human rights record of their nation?

It is obvious that tolerance of other cultural and political systems must stop short of the violation of absolute moral norms—as a consequence, a violation of human rights (such as apartheid) should not be tolerated with the pretext of cultural or ethical relativism. As a private individual, I share Sir Geoffrey Chandler's belief that the days when companies could remain silent about human rights issues are over:

"Silence or inaction will be seen to provide comfort to oppression and may be adjudged complicity.... Silence is not neutrality. To do nothing is not an option."[31]

I am also aware of the dilemma that on the one hand industry associations or chambers of commerce are stronger because they are collective voices for corporate human rights lobbying, but on the other hand associations

"... too often adopt a lowest-common-denominator approach to human rights issues, doing as much in the human rights sphere as their least courageous members (i.e. often nothing at all)." (op cit, p. 24)

It would be an encouraging first step if managers would give the right signals to human rights violators at non-official events by refusing to rationalize what cannot be rationalized, by refusing to level down what should not be leveled down, and by not trivializing what is not at all trivial to the victims of the human rights violations. Martin Luther King Jr. left us with the legacy that

"We shall have to repent in this generation, not so much for the evil deeds of the wicked people, but for the appalling silence of the good people."

What Constitutes an Appropriate Verification Process
for Corporate Human Rights Performance?

Christopher Avery reminds us that twenty years ago most people probably would have given business the benefit of the doubt in a human rights controversy, but that this is no longer the case:

"In the past two decades they have been disappointed too many times by disclosures about the human rights record of particular companies. While they welcome news that a company has adopted a human rights policy, they now withhold judgment to see whether the company follows through with action and whether the results have been verified by an organization truly independent of the company and without any motive to sugar-coat the findings." (op cit, p. 23)

For this reason, companies propose verification processes by external auditors along the same lines as done to audit the companies' books. Following the scandals of Enron and others, representatives of NGOs are skeptical of such solutions. They see a risk that auditors who are profitably associated with the company in other business areas are not 'credible third parties'. For human rights verifications, for instance, they are probably not prepared to jeopardize their main business (auditing of the books) through a critical reporting on possible deviations from the path of virtue.

The 'general overall principles of independent monitoring' for claims of good employer practices developed by Elaine Bernard, Director of the Harvard Trade Union Program, offer a good base of reference for external verification of a corporate human rights performance.[32] Credible human rights verifications must be

– **Independent** from the business enterprise being monitored. This independence, however, should not be defined so restrictively that the corporation monitored is not allowed to pay for it—this is not sustainable for any monitoring institution;

– **Ongoing**, that is, according to a plan being announced on a relatively short-term base and not simply a superficial 'celebrity' visit. All parties affected by a human-rights-relevant business activity must be able to talk with monitors in complete confidentiality and without reprisals;

– **Institutional**, in the sense that the monitoring agency must have independent authority and sufficient resources;

– **Indigenous** where local indigenous people are affected, the monitoring process must include people who speak the language and live in the country where the human rights performance is being monitored;

– **Trusted**, that is, with a track record within the area of competence;

– **Knowledgeable** about the business activities under review and with an appreciation of what is common practice and what is not;

– **Transparent**, that is, as open as possible and—after giving the monitored corporation an opportunity to comment on and, if necessary, initiate necessary steps to correct deficits—with the right to communicate information without corporate prescreening or control.

The monitoring system developed and implemented in the context of the Mattel toy company by S. Prakash Sethi's Monitoring Council, in which the claims that a company has made voluntarily and publicly are the focus of monitoring, holds a lot of promise for serious companies who 'walk as they talk.' (Sethi 2003)

So who would be suitable candidates as monitors? Theoretically it would be ideal if specialized institutions that enjoy high levels of authority and credibility, such as Amnesty International or Human Rights Watch, could take on such verifications on behalf of companies. To do this with any sustainable success, however, it would on the one hand have to be possible for companies to pay for these services, just as it is the case with financial auditing companies. On the other hand, it would have to be ensured that a verification process takes place in a way that both parties could consent to. A company that systematically puts publicly proclaimed values into practice in the form of corporate citizenship guidelines and lives up to these in the form of consistent business practices has little to fear from external verifications.

However, it is in the nature of human beings that they tend to commit individual lapses, make stupid mistakes, or get priorities wrong. Every company with more than 1,000 employees must expect a normal distribution of individual virtues, social competence, or other elements of the collective profile of strengths and weaknesses—and thus also individual misconduct. Where corresponding management processes are implemented and such misconduct can be uncovered and corrected as a result, it can at least be shown that individual problem cases by no means represent 'company policy'. In this respect, too, the presumption of innocence should apply—or at least there should be a certain 'power sharing'. It cannot be that the functions of 'police', 'prosecutor', and 'judge' are in the same hands in the case of external verifications.

Where a deviation from the rule is the exception and not the norm but external verification reports fail to make this transparent, even best practice companies will seek alternative processes. Activist groups who engage in naming and shaming with generalizing preconceptions are understanda-

bly not at the top of the wish list of companies seeking verification services—but the dislike is probably reciprocal. With institutions on record
for activism, there is at least a risk that there is a self-imposed compulsion
to detect some form of misconduct in order to retain credibility with their
own constituency—and then perhaps to make a 'molehill' of individual
laxness into the 'mountain' of a business policy that violates human rights.

What are Useful Indicators?

An important step toward the acceptance of verification processes is a
broad-based agreement on practicable human rights indicators—for example, in collaboration with the Global Reporting Initiative (GRI). Indicators
help a company to translate their commitment to human rights into tangible
and concrete human rights 'deliverables'. At the same time they communicate to the outside world and to human rights NGOs what responsibilities
the company is willing to fulfill and in what way. There are no harmonized
expectations and thus there will not be consensus among human rights
stakeholders on all indicators, and so the debate will continue—but on a
higher and better-informed level.

The general human rights indicators proposed by the GRI (see Table 1)
encompass all essential problems of relevance to business and therefore offer a good approach to gaining a *general* picture of a company's human
rights performance.

To prevent unnecessary and high administrative costs, duplication of effort must be avoided. Unlike in the current 'business and human rights'
debate, no new indicators should be created for 'labor standards' and 'environment', but those already in place should be drawn on (the GRI, for
example). After reaching agreement on a selection of these indicators (HR
5, 6, and 7 might be taken care of by labor standards, while HR 11 and 14
might not be relevant to all sectors), a deeper sector-specific workup is
needed, because factors of importance for different industries differ.

Due to the dimension of human rights deficits and the characteristics of the
different generations of rights ('freedom from torture' is a first-generation
right, for instance, while 'right to medical care' is a second-generation right),
it is further advisable to differentiate the indicators into 'respecting', 'protecting', and 'fulfilling' rights dimensions.[33] Companies must respect human
rights in the sense of refraining from interfering with people's pursuit of their
rights and they must—to the best of their abilities—protect in the sense of
preventing violations by other actors. But they can only to a limited degree fulfill, for example, economic, social, and cultural rights when and
where the primary bearer of duty, the state, is not able or willing to do

that. Philanthropic efforts, as laudable as they may be in fulfilling economic, social, or cultural rights, will not compensate for non-compliance with human rights essentials in normal business activities (such as benefiting from child labor).

Table 1. Indicators for general human rights performance

Core Indicators	Additional Indicators
HR 1 Description of policies, guidelines, corporate structure, and procedures to deal with all aspects of human rights relevant to operations, including monitoring mechanisms and results.	HR 8 Employee training on policies and practices concerning all aspects of human rights relevant to operations.
HR 2 Evidence of consideration of human rights impacts as part of investment and procurement decisions, including selection of suppliers/ contractors.	HR 9 Description of appeal practices, including but not limited to human rights issues.
HR 3 Description of policies and procedures to evaluate and address human rights performance within the supply chain and contractors, including monitoring.	HR 10 Description of non-retaliation policy and effective, confidential employee grievance system (including but not limited to its impact on human rights).
HR 4 Description of global policy and procedure and programs preventing all forms of discrimination in operations, including monitoring.	HR 11 Human rights training for security personnel.
HR 5 Description of freedom of association policy and extent to which this policy is universally applied independent of local laws.	HR 12 Description of policies, guidelines, and procedures to address the needs of indigenous people.
HR 6 Description of policy excluding child labor as defined by ILO Convention 138 and the extent to which this policy is visibly stated and applied, as well as description of procedures and programs to address this issue, including monitoring.	HR 13 Description of jointly managed community grievance mechanisms or authority.
HR 7 Description of policy to prevent forced and compulsory labor and the extent to which this policy is visibly stated and applied, as well as description of procedures and programs to address this issue, including monitoring.	HR 14 Share of operating revenues from the area of operations that is distributed to local communities.

Source: Global Reporting Initiative (2002)

How Can We Develop a Human Rights-Related 'Richter Scale'?

If the media report on the 'human rights abuses' of a company, it is highly likely that concerned people will associate these with very severe human rights abuses. People are likely to think of 'complicity in the abuses of foreign governments related to genocide, war crimes, slavery, torture, exe-cution, crimes against humanity or unlawful detention'.[34] In reality, how-ever, it may be that the facts in question are an 'abuse of human rights' only from the point of view of specific, individual preferences. Human rights violations can be in the eye of the beholder, and the public debate re-flects different views, all of them with merit. But if the critical view is situated well away from the mainstream of the debate, it will not represent a relevant benchmark for a company—still, the damage to the reputation is done.

The current spectrum of the discussion on human rights and business is extremely broad, covering questions of free trade and investment (UN-ECOSOC 2003) as well as bioethical issues concerning the human ge-nome[35] and research priorities of the pharmaceutical industry (Swithern 2003). If we assume that violations of the right to life, slave labor, or child labor represent a different 'quality' of human rights abuses compared, for example, with the profit focus rather than the poverty focus of the research priorities of a pharmaceutical company, then it becomes necessary to dif-ferentiate between varying degrees of human rights violations.

A good simile for what is sought here is the 'Richter scale'. Earthquakes are measured on the Richter scale, by which even people untrained in seismology can approximately estimate the severity of an earthquake. But what about human rights abuses? Are we talking here about research with embryonic stem cells, to which people attribute the whole potentiality of a human being on the basis of their religious beliefs or value systems, and thus also rights and a dignity that are capable of being abused? Or are we talking about contempt for humanity as manifest, for example, in child la-bor in gold and diamond mines of poor countries? Does the severity of the violation in the various cases not differ enormously, and should this not be taken into account with an appropriate weighting? Even given acceptance of the 'universality, indivisibility, interdependence, and interrelatedness of all human rights', it is clear that at the level of socioeconomic develop-ment, different rights carry different degrees of weight. For example, there can be no legitimacy for 'torture' or 'political murder' at any level of de-velopment (a '10' on my human rights measurement scale), but the 'right to rest and leisure, including reasonable limitations of working hours and periodic holidays with pay' (UDHR Article 24), on the other hand—while important—comes under a less essential category (a '2' on my scale).

Through general indicators such as those of the GRI in conjunction with sector-specific indicators, it should be possible to take the political heat out of human rights reporting and to depoliticize the verification process.

Through grouping and weighting of different indicators, it then becomes possible to draw distinctions, such as:

- 'Code Green', which refers to lesser sins of omission that can be easily remedied: for instance, pregnancy tests among women working in production (Article 12 'interference with privacy') or regular overtime among members of management (Article 24 'right to leisure');
- 'Code Orange', such as unknowingly violating but not making efforts to find out; and
- 'Code Red', the systematic violation within corporate activities or direct benefit from violation of subcontractors or subsidiaries.

In his first lecture series in Germany following the end of Nazi power, in the winter term of 1945–46, the German philosopher Karl Jaspers (2000) reflected on 'The Question of German Guilt.' And he identified four types of guilt, which are also of relevance for the present discussion—the following is quoted from the English version:

- **Criminal guilt**: Crimes are acts capable of objective proof and violate unequivocal laws. Jurisdiction rests with the court, which in formal proceedings can be relied upon to find the facts and apply the law;
- **Political guilt**: This, involving the deeds of statesmen and of the citizenry of a state, results in my having to bear the consequences of the deeds of the state whose power governs me and under whose order I live. Everybody is co-responsible for the way he is governed. Jurisdiction rests with the power and the will of the victor, in both domestic and foreign politics. Success decides. Political prudence, which takes the more distant consequences into account, and the acknowledgement of norms, which are applied as natural and international law, serves to mitigate arbitrary power;
- **Moral guilt**: I, who cannot act otherwise than as an individual, am morally responsible for all my deeds, including the execution of political and military orders. It is never simply true that 'orders are orders'. Rather—as crimes even though ordered (although, depending on the degree of danger, blackmail and terrorism, there may be mitigating circumstances)—so every deed remains subject to moral judgment. Jurisdiction rests with my conscience and in communication with my friends and intimates who are lovingly concerned about my soul;
- **Metaphysical guilt**: There exists a solidarity among men as human beings that makes each co-responsible for every wrong and every injustice in the world, especially for crimes committed in his presence or with his

knowledge. If I fail to do whatever I can to prevent them, I too am guilty. If I was present at the murder of others without risking my life to prevent it, I feel guilty in a way not adequately conceivable either legally, politically or morally.... jurisdiction rests with God alone.

With these distinctions between different types of guilt, Jaspers (2000) sought to

"... preserve us from the superficiality of talk about guilt that flattens everything out on a single plane, there to assess it with all the crudeness and lack of discrimination of a bad judge."

Enlightened corporations, under all circumstances, will shy away from criminal guilt; they will create a corporate governance structure to avoid the political guilt of not making unmistakably clear what are the corporate 'do's' and the 'dont's' and, last but not least, strive for a management who feels also morally accountable for actions and omissions. As long as Amnesty International annual reports have more than 20 pages, we all will have to live with metaphysical guilt.

Preliminary Conclusions

The *Universal Declaration of Human Rights* represents the most important value catalogue for human beings in all cultures and at all times. This declaration affirms that there are certain non-negotiable rights that are enjoyed by all people in all places at all times based simply on the fact that they are human beings. It is precisely in the context of globalization, where different cultures, social constitutions, and socioeconomic conditions meet, that this common denominator is also of utmost importance to companies. Business enterprises need to do their respective 'homework' and act consistently in order to adjust the corporate social responsibility concept to the changed sociopolitical framework of a globalizing world (Leisinger and Schmitt 2003).

Beyond the day-to-day responsibilities of business, one of the most important questions for managers of global companies is 'What kind of a world do we wish for ourselves and our children?' Whatever the individual value-based preferences may be regarding a right to life in dignity, justice, equality of opportunity, and fairness, it cannot be a world in which human rights are not respected. And what duties are we prepared to assume to ensure that our vision of an 'ideal' world as we see it can be achieved? This is something that has to be decided by every individual in their families, in their jobs, and in their role as citizen. Just as in elections, anyone can say

"... that if he does not vote, it will not change the election result, but he will vote anyway because he knows that all individuals together make up the result. So the moral force of the seemingly vanishing individual is the only substance and the true factor for what becomes of humanness." (Jaspers 1949)

Change—even corporate change—is always initiated by minorities, by intellectual elites who take the risk upon themselves of being pioneers in uncharted territory; 'the big values always remain closely tied to the small number' (Guardini 1986). All those who are making their contribution to the world they would like to see for their children should be confident, for as Margaret Mead once reminded us,

"Never doubt that a small group of thoughtful, committed citizens can change the world; indeed it's the only thing that ever has."[36]

Endnotes

[1] Sir Geoffrey Chandler from his foreword to Amnesty International/The Prince of Wales Business Leaders Forum (2000) p. 5

[2] For examples of current problems, see Amnesty International (2003) and Amnesty International and Democratic Republic of Congo (2003); see also //action.web.ca/home/pac/attach/w_africa_e.pdf and Gberie (2003); see also Chandler (1998).

[3] Anyone who enters the two terms in an Internet search engine such as Google will find over 5 million contributions to the debate.

[4] See the data of Globe Scan Research Teams/Environics International Ltd., Toronto 2003.

[5] See www.unglobalcompact.org.

[6] See also the accompanying Commentary, UN Doc.E/CN.4/Sub.2/2003/38/Rev.2.

[7] Whereas the UN Global Compact is a voluntary framework for promoting good corporate citizenship, the UN Norms are seen to become a part of international law and perceived to suggest that companies should be subject to the kind of enforcement by the UN Commission for Human Rights that has previously applied only to nation states. Another point to be discussed in more detail in the negotiation process ahead is the "periodic monitoring and verification by the United Nations and other international and national mechanisms already in existence or yet to be created." See Amnesty International's contribution to this debate (2004).

[8] This sentence conceals an even greater complexity, which has best been explained in Niklas Luhmann (1997).

[9] See e.g., the report of the Novartis Foundation for Sustainable Development, Basel 2003 or www.novartisfoundation.com.

[10] See www.unhchr.ch/Udhr

[11] Art. 25 UDHR as well as Article 11 and 12 of the International Covenant on Economic, Social and Cultural Rights.

[12] Discrimination means any distinction, exclusion, or preference made that has the effect of nullifying or impairing equality of opportunity or treatment in employment or occupation.

[13] For example, the UN Basic Principles on the Use of Force and Firearms.

[14] As forbidden, for example, in ILO Conventions 29 and 105.

[15] Economic exploitation of children includes employment or work in any occupation before a child completes compulsory schooling and, in any case, before the child reaches 15 years of age. Economic exploitation also includes the employment of children in a manner that is harmful to their health or development or will prevent the children from attending school or performing school-related responsibilities. Economic exploitation does not include work done by children in schools for general, vocational or technical education, or in other training institutions.

[16] For example, as those found in the International Covenant on Economic, Social and Cultural Rights and the respective ILO Conventions.

[17] See Sachs (2003) for conflict situations between resource use and subsistence rights.

[18] See the annual reports of Amnesty International.

[19] See e.g., *New Academy Review*, Vol. 2, No. 1 (Spring 2003): "Business interests ... have been antagonistic to human rights" (p. 50) or "MNCs can now pose a significant threat to human rights, and also undermine the ability of individual states to protect people from human rights abuses" (p. 92).

[20] The text of the preamble says "that every individual and every organ of society, keeping this Declaration constantly in mind, shall strive by teaching and education to promote respect for these rights and freedoms and by progressive measures, national and international, to secure their universal and effective recognition and observance."

[21] The UN Global Compact, too, so far enjoys the support of fewer than 1,200 of the 70,000 or more companies with international operations.

[22] See Duesenberry (1967); for a short introduction see cepa.newschool.edu/het/essays/multiacc/ratchet.htm

[23] To support this notion see also Avery (2000).

[24] See www.unglobalcompact.org; see also Amnesty International and The Prince of Wales Business Leaders Forum (2000) pp. 28ff.

[25] In the UN Global Compact Resource Package – Human Rights Presentation.

[26] See www.amnesty.it/edu/formazione/mondo_economico/mitw/documenti/Corporate_complicity.doc; see also Stoett (2002).

[27] www.rachel.org/bulletin/bulletin.cfm?Issue_ID = 532

[28] I am grateful to Paul Streeten for his comments on the precautionary principle.

[29] See Amnesty International and The Prince of Wales Business Leaders Forum (2000) pp. 30ff.

[30] After the 1973 coup d'état of the Chilean military against President Salvador Allende, the suspected support of the US-American ITT corporation led to

widespread protests and to a UN General Assembly Resolution (May 1, 1974) calling for an international Code of Conduct preventing interference with the 'internal affairs' of the countries within which companies operate (ECOSOC-Commission for Transnational Corporations: Material Relevant to the Formulation of a Code of Conduct, 10 December 1976, §59). This view was confirmed by the UN Charter of Economic Rights and Duties and taken up by the 1976 version of the OECD Guidelines for Multinational Corporations. See: United Nations Division on Transnational Corporations and Investment (1996).

[31] Chandler (1997) quoted from Avery (2000) p. 22.

[32] Bernard (1997) quoted in Avery (2000) p. 51.

[33] As UNDP (2000) did in its *Human Development Report.*

[34] Mary Robinson points to such issues in Sullivan (2003).

[35] For example The Universal Declaration on the Human Genome and Human Rights. portal.unesco.org/en/ev.php-URL_ID = 13177&URL_DO = DO_TOPIC &URL_SECTION = 201.html

[36] I owe this quote to a poster shown at the 2003 Business and Human Rights seminar of the "Business Leaders Initiative on Human Rights" (Honorary Chair: Mary Robinson), London, 9 December 2003.

References

Amnesty International and The Prince of Wales Business Leaders Forum (2000) *Human Rights—Is It Any of Your Business?* Amnesty International, London

Amnesty International (2003) *Human Rights on the Line. The Baku-Tbilisi-Ceyhan Pipeline Project.* Amnesty International, London

Amnesty International and Democratic Republic of Congo (2003) *Our Brothers Who Help to Kill US—Economic Exploitation and Human Rights Abuses in the East.* Amnesty International, London

Amnesty International (2004) *The UN Human Right Norms for Business: Towards Legal Accountability.* Amnesty International, London

Avery CL (2000) *Business and Human Rights in a Time of Change.* Amnesty International, London

Bernard E (1997) Ensuring Monitoring is Not Co-opted. *New Solutions*, Vol. 7, No. 4

CETIM and American Association of Jurists (2002) *Will the UN Compel Transnational Corporations to Comply with International Human Rights Standards?* Geneva New York, pp. 10f

Chandler G (1998) Oil Companies and Human Rights. *Business Ethics. A European Review*, Vol. 7, No. 2, pp. 69–72

Clapman A and Jerbi S (2001) Categories of Corporate Complicity in Human Rights Abuses. *24 Hastings International and Comparative Law Review*, pp. 339–349

Davis R and Nelson J (2003) *The Buck Stops Where? Managing the Boundaries of Business Engagement in Global Development Challenges.* International Business Leaders Forum, London, p. 3

de George RT (1993) *Competing with Integrity in International Business.* Oxford University Press, Oxford New York

Donaldson T and Dunfee TW (1999) *Ties That Bind. A Social Contracts Approach to Business Ethics.* Harvard Business School Press, Boston

Drucker P (1993) *Post-Capitalist Society.* Harper Business, New York

Duesenberry JS (1967) *Income, Savings, and the Theory of Consumer Behavior.* Oxford University Press, New York

Gberie L (2003) *West Africa: Rocks in a Hard Place.* The Diamonds and Human Security Project. Occasional Paper, No. 9, Ottawa

Global Reporting Initiative (2002) *Sustainability Reporting Guidelines.* Boston, pp. 53/54. www.globalreporting.org/guidelines/2002/contents.asp

Guardini R (1986) *Das Ende der Neuzeit. Die Macht.* Grünewald/Schöningh, Mainz Würzburg

International Chamber of Commerce and International Organization of Employers (2003) *Joint views of the IOE and ICC on the Draft Norms on the Responsibilities of Transnational Corporations and Other Business Enterprises With Regard to Human Rights,* 22 July 2003. Paris Geneva

Jaspers K (1949) Über Bedingungen und Möglichkeiten eines neuen Humanismus. *Die Wandlung.* Herbstheft. Schneider Verlag, Heidelberg, p. 734

Jaspers K (2000) *The Question of German Guilt.* Fordham University Press, New York

Leisinger KM (2003) The Benefits and Risks of the UN Global Compact: The Novartis Case Study. *The Journal of Corporate Citizenship,* Autumn 2003, pp. 113–131

Leisinger KM and Schmitt K (2003) *Corporate Ethics in a Time of Globalization.* Sarvodaya Vishva Lekha Publishers, Colombo, Sri Lanka

Litvin D (2003) Raising Human Rights Standards in the Private Sector. *Foreign Policy,* November/December 2003, pp. 68–72

Luhmann N (1997) *Die Gesellschaft der Gesellschaft.* Suhrkamp Verlag, Frankfurt am Main

Montague P (1998) "The Precautionary Principle." *Rachel's Environment and Health Weekly,* Vol. 586, 19 February 1998. www.rachel.org/bulletin/bulletin.cfm?Issue_ID = 532

OECD (2000) *The OECD Guidelines for Multinational Enterprises: Revision 2000.* Paris. http://www.oecd.org/dataoecd/56/36/ 1922428.pdf

Sachs W (2003) *Environment and Human Rights.* Wuppertal Institute for Climate, Environment and Energy, Wuppertal

Schrage E (2003) Emerging Threats: Human Rights Claims. *Harvard Business Review,* August 2003. Cambridge MA, pp. 16ff

Sethi SP (2003) *Setting Global Standards: Guidelines for Creating Codes of Conduct in Multinational Companies.* Wiley and Sons, New York

Stoett P (2002) Shades of Complicity: Towards a Typology of Transnational Crimes Against Humanity. Jones A (ed) *Genocide, War Crimes, and the West: Ending the Culture of Impunity.* Zed Books, London

Streeten P (2003) *Should Companies Try to Do Good?* Background paper contribution to the "Human Rights and Business" debate of the Novartis Foundation for Sustainable Development, February 2003

Sullivan R ed (2003) *Business and Human Rights. Dilemmas and Solutions.* Greenleaf Publishing, Sheffield

Swithern S (2003) From Bhopal to Doha: Business and the Right to Health. *New Academy Review*, Vol. 2, No. 1, pp. 50–61

UN Sub-Commission on the Promotion and Protection of Human Rights (2003) *Norms on the Responsibilities of Transnational Corporations and Other Business Enterprises With Regard to Human Rights.* E/CN.4/sub.2/2003/12/Rev.2, 26 August 2003. Geneva

UNDP (2000) *Human Development Report 2000. Human Rights and Human Development.* Oxford University Press, Oxford New York

UN-ECOSOC (2003) *Human Rights, Trade and Investment.* E/CN.4/Sub.2/2003/9, 2 July 2003

UN-ECOSOC Commission for Transnational Corporations (1976) *Material Relevant to the Formulation of a Code of Conduct,* 10 December 1976, p. 59

A New Glasnost for Global Sustainability[1]

MIKHAIL GORBACHEV[2]

It may seem paradoxical, but despite having borne witness to the countless humanitarian and environmental disasters of the past decades, I am still an optimist. After all, we have been exposed to an avalanche of grim forecasts regarding our future, seemingly leaving little, if any, room for optimism. But being an optimist does not mean simply looking at the world through rose-tinted glasses, like Voltaire's Candide, and declaring everything to be for the best despite an endless array of misfortunes. Being an optimist, as I see it, means to refuse to make do with the status quo and instead to consciously look for ways to make the world a better place and help address the practical challenges faced by people here and now. I call this 'optimism by action', and I believe that such a philosophy of life could provide the catalyst for the much-needed transformation to sustainable development. The first step is to inform and motivate the people.

Twenty years ago, when the idea of *Glasnost*, or openness, was used to launch the process of *Perestroika* that transformed the Soviet Union, no one believed that it was doable. But I was driven by the need to 'wake up' those people who had 'fallen asleep' and make them truly active and concerned, to ensure that everyone felt as if they were master of the country, of their enterprise, office or institute—to get the individual involved in all processes. One of the first outcomes of Glasnost in the USSR was heightened awareness of the massive environmental problems blighting the country and impassioned public demands to stop the most damaging activities, resulting in the closure of thousands of heavy polluting factories and cancellation of a major project to divert Siberian rivers.

Today I am convinced that the citizens of the world need a reformulated Glasnost to invigorate, inform and inspire them to put the staggering resources of our planet and our knowledge to use for the benefit of all citizens of the earth and not go back to the days of prolific military spending and fear of people whose ways are different from our own. People cannot

M. Keiner (ed.), The Future of Sustainability, 153–160.

tolerate living on a planet where millions of children have no clean water to drink and go to sleep hungry, once they know that they have the power to change it. I have faith in humankind, and it is this faith that has allowed me to remain an active optimist.

As the stakes rise higher, with the permanent damage we are doing to our planet and the erosion of global security, there is no time to be lost in addressing the three principle and inter-linked challenges of sustainable development: peace and security, poverty and deprivation, and the environment. In the face of international terrorism, the threat of proliferation of weapons of mass destruction and frequent local armed conflicts, continuous efforts are needed to ensure peace and security. The existence of enormous poverty-stricken areas in the world is morally unacceptable and provides the breeding grounds for extremism, violence and organized crime unconstrained by any borders. The global environment displays alarming signs of discontent, and its problems are no longer localized and manageable. Our damage to the Earth's atmosphere is causing our climate to change, natural disasters to become more frequent and devastating, glaciers to melt, and the polar ice caps to thin; this is coupled with the results of irresponsible business practices, where ocean fish stocks are depleting, deserts are advancing and thousands of plant and animal species continue to disappear.

We are risking our future for an ephemeral, pollution and exploitation based prosperity. Disaster, in the form of an oil spill, chemical leak or even nuclear accident like Chernobyl, could strike any day with little being done in the way of prevention. In order not to let this happen, we must put an end to the conspiracy of silence of those who are unwilling to change their lifestyles or risk disturbing the foundations of the economic system that pays their bills, and expose the terrible moral cowardice of those politicians who cover this conspiracy up, refusing to recognize the true extent and nature of modern challenges.

There are clear links between the three sustainable development challenges, both in terms of origin, repercussions and the imperatives they dictate to humankind. One cannot counteract bigotry, crime and terrorism or ensure global security without combating poverty. One cannot address poverty without protecting our human right to fulfill our basic needs, and ensuring both environmental protection and equal access to natural resources for all. Human development and environmental protection are interdependent objectives. How can one tell the poor of the Amazon basin not to chop down the rainforest to lay a field if this is their only means of making ends meet? How can one demand cost-prohibitive environmental protection measures from a poor country? On the other hand, if nature is not given enough consideration, our efforts to build a more equitable and better world are doomed to founder.

Reflecting upon this, one cannot help wondering what caused the situation we now have on our hands. If the causes are unclear, no rational solution is possible. Our world is becoming rife with conflicts and contradictions, problems with a long heritage whose preconditions have been amassing during the evolution of human civilization. Today, these conflicts have reached truly global proportions and stand to jeopardize the basic security of humankind. And globalization, as the prevailing force in world development, must be held responsible. Globalization lays bare and intensifies all the conflicts and contradictions of the past and drives them to dangerous degrees.

The world's market-driven globalization tends to enforce the notion, derived from neo-liberal theory, that economic growth as measured by GNP/GDP indicators is the only way to measure national wealth and progress. Capital accumulation and individual consumption are given a higher status than social and spiritual values or cultural heritage. Ideology and policies of the neo-liberal globalism initiated by the countries that have benefited most from globalization make this trend that much stronger. The cumulative results of all the individual decisions based on this logic in the long run lead to unforeseen and dangerous consequences.

One often comes across the argument that globalization, as we know it, is a *fait accompli*, a process entirely outside our control. Particularly vociferous with this argument, unsurprisingly, are those who want to instill in the public mind the futility and pointlessness of any opposition to globalization. In the meantime, several highly influential researchers have argued convincingly about the role of political choices as a factor used to harness globalization and make it work for the benefit of major players on the global marketplace.

That politics lies behind globalization is unquestionable. In recent years this has been clearly illustrated by the neo-conservatives in the USA seeking to take advantage of globalization to pursue an imperialist policy of force and impose their will upon the rest of the world. The reason why the force factor comes to the fore possibly lies in the realization of a few simple facts: natural resources are finite, their use has already exceeded a critical point, and the capture of the lion's share by a smaller (and decreasing) portion of the humanity deprives the rest of the world (and growing majority) of equal access to such resources and, in many cases, to the essential means of subsistence.

Is there an alternative to the existing situation? It is my conviction that our history is not predetermined and there is room for an alternative in any situation. It was this pursuit of an alternative development model that led to the elaboration of a sustainable development program for the world. Agenda 21 was supported by the United Nations and endorsed by the heads of state and government of most states in 1992 (UN 1992). For the first time in history, the world community managed to map out and agree a general strategic plan designed to address people's vital problems. However, serious obstacles emerged as implementation started. The governments of industrialized countries chose to retract from their commitments, in particular regarding the increase of their development aid, in favor of the philosophy of economic liberalism, deregulation and accelerated economic growth.

Opponents of the sustainable development paradigm have spared no effort in trying to discredit the idea in the public mind. And yet, the interest is still there. The so-called 'anti-globalization movement' (in effect, it is the movement against market-driven fundamentalism) is in favor of an alternative development model. Their motto is, 'Another World is Possible'. International social democracy, rural people, 'green' movements worldwide, and thousands of NGOs representing millions of members, also stand behind the sustainable development principle. We are talking about a powerful force whose pressure is being increasingly felt by the ruling elite.

So what can we do to make a difference? First of all, we need to scrutinize the structural factors inhibiting the transition to sustainable development. We need to better understand the mechanisms of globalization that are directing development on such a dangerous course. We need to bridge the gap between our moral consciousness and the challenges of time. Consumerism and national egocentrism continue to pose a serious threat to achieving sustainable development goals. A turnaround will not be possible unless the breach between the objective need to reverse currently prevalent behavioral patterns and the subjective unwillingness of states, communities and individuals to do so is overcome. This turnaround must begin with changes in the human spirit and a reprioritization of our value system, including relations between people and the human-nature interrelationship.

Presently, politics lags behind the pace of change. Greater analysis of global issues and corresponding recommendations to politicians are needed, and hence the role of science and education ought to be enhanced. There is an urgent need for environmental education and the respect for nature. All this gives science, education and the mass media a special role and responsibility. Prominent scientists have been cautioning us about the dangers looming over humankind for many years: Sadly, they have been paid little heed to, often ignored, and even forgotten altogether. Facts about global threats gathered by scientists should become public knowledge, and the main vehicle for translating scientific conclusions into terms we can all understand is the mass media. The mass media is needed to build a bridge between civil society and political and economic leaders. The antiglobalization movement simply says that a different world is possible, and the Socialist International adopted the idea of 'globalization with a human face' from the United Nations Human Development Report (1999). These attitudes and initiatives on their own do not promise any real change. Thus, the media has an exceptionally important role in building a 'society of knowledge'. Interaction of science and mass media is becoming crucial today. The more the society relies on true knowledge, the more desperately it needs it.

But the media is not always consistent; it is often controversial and, sometimes, even counterproductive. Scientists and the media both suffer from a 'credibility gap'. Too often, the media less informs viewing, reading or listening audiences, as it misleads them. It makes use of cheap sensationalism to satisfy raw tastes and thus manipulate the public mind. Scientists themselves could use a little help coordinating their actions and getting the truth of their findings across.

Apart from science and the media, the education system is another essential channel of knowledge regarding global challenges and sustainable development. Practically every activity in our times requires knowledge in the area of environmental protection. It is important that people learn, starting in school, how to respect nature, save energy and water resources, and manage domestic waste. Schools of all levels are called upon to instill the concepts of human togetherness, world integrity and the culture of solidarity and peace in their students.

Glasnost could be put to service as a catchall phrase for all of these weapons in the struggle for transparency and awareness. Glasnost is more than transparency; it is a demanding, long-term process of awakening, which inevitably lead to calls for fundamental changes. In the field of sustainable development, such a process is needed to combat apathy, to engage the people to the task of choosing more equitable and sustainable lifestyles, and to address the dominance of short-term interests and lack of transparency at the decision-making level. A process of Glasnost would tackle both aspects of this dangerous blend of indifference and concealment and ultimately rebuild the trust between people, business and government, desperately needed if we are going to stand any chance in achieving the Millennium Development Goals to combat poverty, disease and deprivation by 2015 (UN 2000a).

The escalation of global problems is in many ways attributable to world politics lagging behind the real processes unfolding in the world. World politics is skidding, proving to be incapable of responding to the challenges of globalization. I am personally enormously disappointed that, more than a decade after it was given a new lease of life with the end of the Cold War, multilateralism is foundering. We have squandered much of the capital of trust and cooperation that emerged at the end of the 20th Century. I am convinced that contemporary world politics is not to be based on the conventional principle of balance of powers, but rather on the balance of interests, and that dialogue between cultures and civilizations must become its primary tool. Politics should concentrate on avenues of cooperation and ways to break through deadlocks by promoting just and long-term real world solutions, not quick fixes or inequitable compromises.

In the past several years, a number of prominent civil and political leaders have gone to great lengths to develop moral frameworks for sustainable development. These efforts bore fruition in the form of the Earth Charter, a code of ethics for the planet (Earth Charter Commission 2000). The Earth Charter outlines the interrelationship between humans and the rest of nature: it spells out a new set of ecological principles as guidelines for human behavior. The Charter has become an important document in the sustainable development field and conferences and outreach are part of the process of promoting an understanding of the 16 principles of the Earth Charter. Today, the Earth Charter is endorsed by more than 8,000 organizations that represent hundreds of millions of people.

Under current circumstances, it is becoming an extremely pressing task to have this code of basic moral principles observed by governments, business and NGOs simply in order give future generations and our planet a chance to survive. In a world increasingly besieged by corruption, greed and self-interest, we need leaders who have the moral courage to ground their decisions in this new global ethic and sustainable development principles.

Among these principles, solidarity takes a special place. The principle of solidarity has played a vital role at all times, especially in small groups, communities and social movements, but in this day and age, the imperative of global solidarity moves to the foreground. This means solidarity of a higher order, to meet the requirements of globalization as the dominant trend of modern world development. It is solidarity that presents itself as the pillar of sustainable development, including all human and intergenerational aspects.

Big business and especially multinational corporations are often blamed for causing or escalating social and ecological problems. There are good reasons for that. Being a part of the preponderant socioeconomic system, business definitely bears the earmarks of the system's known ills, e.g., frequent scandalous violations of ethics and corruption. At the same time, by abiding by appropriate codes of ethics, business could play a fundamental part in the protection of the environment and combating poverty, as is already being demonstrated by some enterprises.

Therefore, the Global Compact proposal initiated by Kofi Annan as a cooperation mechanism between the United Nations and private business for addressing development problems is certainly worth supporting (UN 2000b). Corporations that join the Global Compact undertake commitments to implement certain arrangements relating to human rights, labor laws and environmental protection, as well as express their willingness to report to the United Nations on a regular basis. The World Summit on Sustainable Development in Johannesburg in 2002 was a landmark in the business of forging partnerships between the United Nations, governments, business and NGOs to pool resources for addressing global environment, health and poverty challenges.

In the Millennium Declaration adopted by the UN General Assembly in September 2000, world leaders reaffirmed their support for sustainable development principles and registered concern over the obstacles that developing countries have to face in trying to mobilize resources for sustainable development funding (UN 2000a). The Declaration underlines the significance of solidarity as one of the essential values relevant for international relations in the 21st Century. The Millennium Development Goals formulated in the Declaration with specific targets and timeframes serve as a specific demonstration of this commitment.

To achieve these development goals and end the growing scourge of poverty and disease, we will first have to address one of the most important problems discussed throughout the world today—global governance and in particular, governance over globalization. In considering the strongest possible governance framework for the global system, several issues must be addressed. There is an imminent tension between the strong and weak nation states, each of which have different priorities for global governance. There are tensions between the North versus the South and then of course the tensions with the US, who defies all forms of multilateralism at a time when the world needs these institutions most. The systematic imposition of the US' will on the rest of the world is counterproductive and reverts international relations back to a cold war climate. Governance must be based on internationally recognized moral precepts, as stated in the Millennium Declaration:

"Only through broad and sustained efforts to create a shared future, based upon our common humanity in all its diversity, can globalization be made fully inclusive and equitable." (UN 2000a)

Beautifully and rightly said, but it is important that these words be put to life. The global public should monitor progress by juxtaposing politicians' words against their deeds. "Judge not by words, but by deeds" should be our mantra. That is precisely why we need a new Glasnost to inspire citizens to become actively involved in the struggle for a better tomorrow. I believe in people and will remain an optimist, but one calling for action and positive change.

Endnotes

[1] This chapter is based on Mr. Gorbachev's article "A New Glasnost for the Planet", published in the first issue of *The Optimist* magazine in April 2004. (www.optimistmag.org).

[2] Mikhail Gorbachev is Chairman of the Board of Green Cross International.

References

Earth Charter Commission (2000) *Earth Charter.* March 2000, Paris.www.earth-charter. org

UN (1992) Agenda 21. *Earth Summit, Rio de Janeiro.* www.un.org/esa/sustdev/documents/agenda21/english/agenda21toc.htm

UN (2000a) General Assembly, *United Nations Millennium Declaration,* 8 September 2000. United Nations, New York

UN (2000b) *United Nations Global Compact,* 26 July 2000. New York. www. un-globalcompact.org

UNDP (1999) *Human Development Report.* Oxford University Press, Oxford, New York

Tools for the Transition to Sustainability[1]

DENNIS L. MEADOWS

"We must be careful not to succumb to despair, for there is still the odd glimmer of hope."

Edouard Saouma (1993)[2]

"Can we move nations and people in the direction of sustainability? Such a move would be a modification of society comparable in scale to only two other changes: the Agricultural Revolution of the late Neolithic and the Industrial Revolution of the past two centuries. Those revolutions were gradual, spontaneous, and largely unconscious. This one will have to be a fully conscious operation, guided by the best foresight that science can provide ... If we actually do it, the undertaking will be absolutely unique in humanity's stay on the Earth."

William D. Ruckelshaus (1989)[3]

We have been writing about, talking about, and working toward sustainability for decades now. We have had the privilege of knowing thousands of colleagues in every part of the world who work in their own ways, with their own talents, in their own societies toward a sustainable society. When we act at the official, institutional level and when we listen to political leaders, we often feel frustrated. When we work with individuals, we usually feel encouraged.

Everywhere we find folks who care about the earth, about other people, and about the welfare of their children and grandchildren. They recognize the human misery and the environmental degradation around them, and they question whether policies that promote more growth along the same old lines can make things better. Many of them have a feeling, often hard for them to articulate, that the world is headed in the wrong direction and that preventing disaster will require some big changes. They are willing to work for those changes, if only they could believe their efforts would make a positive difference. They ask: What can I do? What can governments do?

M. Keiner (ed.), The Future of Sustainability, 161–178.

What can corporations do? What can schools, religions, media do? What can citizens, producers, consumers, parents do?

Experiments guided by those questions are more important than any specific answers, though answers abound. There are 'fifty simple things you can do to save the planet'. Buy an energy-efficient car, for one. Recycle your bottles and cans, vote knowledgeably in elections—if you are among those people in the world blessed with cars, bottles, cans, or elections. There are also not-so-simple things to do: Work out your own frugally elegant lifestyle, have at most two children, argue for higher prices on fossil energy (to encourage energy efficiency and stimulate development of renewable energy), work with love and partnership to help one family lift itself out of poverty, find your own 'right livelihood', care well for one piece of land; do whatever you can to oppose systems that oppress people or abuse the earth, run for election yourself.

All these actions will help. And, of course, they are not enough. Sustainability and sufficiency and equity require structural change; they require a revolution, not in the political sense, like the French Revolution, but in the much more profound sense of the Agricultural or Industrial Revolutions. Recycling is important, but by itself it will not bring about a revolution.

What will? In search of an answer, we have found it helpful to try to understand the first two great revolutions in human culture, insofar as historians can reconstruct them.

The First Two Revolutions: Agriculture and Industry

About 10,000 years ago, the human population, after eons of evolution, had reached the huge (for the time) number of about ten million. These people lived as nomadic hunter-gatherers, but in some regions their numbers had begun to overwhelm the once-abundant plants and game. They took two strategies to adapt to the problem of disappearing wild resources: some of them intensified their migratory lifestyle. They moved out of their ancestral homes in Africa and the Middle East and populated other areas of the game-rich world.

Others started domesticating animals, cultivating plants, and *staying in one place.* That was a totally new idea. Simply by staying put, the proto-farmers altered the face of the planet, the thoughts of humankind, and the shape of society in ways they could never have foreseen.

For the first time it made sense to own land. People who didn't have to carry all of their possessions on their backs could accumulate things, and some could accumulate more than others. The ideas of wealth, status, inheritance, trade, money, and power were born. Some people could live on excess food produced by others. They could become full-time toolmakers, musicians, scribes, priests, soldiers, athletes, or kings. Thus arose, for better or worse, guilds, orchestras, libraries, temples, armies, competitive games, dynasties, and cities.

As its inheritors, we think of the Agricultural Revolution as a great step forward. At the time it was probably a mixed blessing. Many anthropologists think that agriculture was not a better way of life, but a necessary one to accommodate increasing populations. Settled farmers got more food from an acre than hunter-gatherers did, but the food was of lower nutritional quality and less variety, and it required much more work to produce. Farmers became vulnerable in ways nomads never were to weather, disease, pests, invasion by outsiders, and oppression from their emerging ruling classes. People who did not move away from their own wastes experienced humankind's first chronic pollution.

Nevertheless, agriculture was a successful response to wildlife scarcity. It permitted yet more population growth, which added up over centuries to an enormous increase, from 10 million to 800 million people by 1750. However, the larger population created new scarcities, especially in land and energy. Another revolution was necessary.

The Industrial Revolution began in England with the substitution of abundant coal for vanishing trees. The use of coal raised practical problems of earth-moving, mine construction, water pumping, transport, and controlled combustion. These problems were solved relatively quickly, resulting in concentrations of labor around mines and mills. The process elevated science and technology to a prominent position in human society— above religion and ethics.

Again everything changed in ways that no one could have imagined. Machines, not land, became the central means of production. Feudalism gave way to capitalism and to capitalism's dissenting offshoot, communism. Roads, railroads, factories, and smokestacks appeared on the landscape. Cities swelled. Again the change was a mixed blessing. Factory labor was even harder and more demeaning than farm labor. The air and waters near the new factories turned unspeakably filthy. The standard of living for most of the industrial workforce was far below that of a farmer. But farmland was not available; work in a factory was.

It is hard for people alive today to appreciate how profoundly the Industrial Revolution changed human thought, because that thought still shapes our perceptions. Historian Donald Worster (1988) has described the philosophical impact of industrialism perhaps as well as any of its inheritors and practitioners can:

"The capitalists ... promised that, through the technological domination of the earth, they could deliver a more fair, rational, efficient and productive life for everyone ... Their method was simply to free individual enterprise from the bonds of traditional hierarchy and community, whether the bondage derived from other humans or the earth ... that meant teaching everyone to treat the earth, as well as each other, with a frank, energetic, self-assertiveness.... People must ... think constantly in terms of making money. They must regard everything around them—the land, its natural resources, their own labor—as potential commodities that might fetch a profit in the market. They must demand the right to produce, buy, and sell those commodities without outside regulation or interference.... As wants multiplied, as markets grew more and more far-flung, the bond between humans and the rest of nature was reduced to the barest instrumentalism."

That bare instrumentalism led to incredible productivity and a world that now supports, at varying levels of sufficiency, 6,000 million people—more than 600 times the population existing before the agricultural revolution. Far-flung markets and swelling demands drive environmental exploitation from the poles to the tropics, from the mountaintops to the ocean depths. The success of the Industrial Revolution, like the previous successes of hunting-gathering and of agriculture, eventually created its own scarcity, not only of game, not only of land, not only of fuels and metals, but of the total carrying capacity of the global environment. Man's ecological footprint had once more exceeded what was sustainable. Success created the necessity for another revolution.

The Next Revolution: Sustainability

It is as impossible now for anyone to describe the world that could evolve from a Sustainability Revolution as it would have been for the farmers of 6000 BC to foresee the corn and soybean fields of modern Iowa, or for an English coal miner of 1750 AD to imagine an automated Toyota assembly line. Like the other great revolutions, though, the coming Sustainability Revolution will also change the face of the land and the foundations of human identities, institutions, and cultures. Like the previous revolutions, it will take centuries to unfold fully—though it is already underway.

Of course no one knows how to bring about such a revolution. There is not a checklist: 'To accomplish a global paradigm shift, follow these twenty steps.' Like the great revolutions that came before, this one can't be planned or dictated. It won't follow a list of fiats from a government or from computer modellers. The Sustainability Revolution will be organic. It will arise from the visions, insights, experiments, and actions of billions of people. The burden of making it happen is not on the shoulders of any one person or group. No one will get the credit, but everyone can contribute.

Our systems training and our own work in the world have affirmed for us two properties of complex systems germane to the sort of profound revolution we are discussing here.

First, information is the key to transformation. That does not necessarily mean *more* information, better statistics, bigger databases, or the World-wide Web, though all of these may play a part. It means *relevant, compelling, select, powerful, timely, accurate* information flowing in new ways to new recipients, carrying new content, suggesting new rules and goals (which are themselves information). Any system will behave differently when its information flows are changed. The policy of *Glasnost,* for example, the simple opening of information channels in the Soviet Union that had long been closed, guaranteed the rapid transformation of Eastern Europe beyond anyone's expectation. The old system had been held in place by tight control of information. Letting go of that control triggered total system restructuring (turbulent and unpredictable, but inevitable).

Second, systems strongly resist changes in their information flows, especially in their rules and goals. It is not surprising that those who benefit from the current system actively oppose such revision. An entrenched system can constrain almost entirely the attempts of an individual or small group to operate by different rules or to attain goals different from those sanctioned by the system. Innovators can be ignored, marginalized, ridiculed, denied promotions or resources or public voices. They can be literally or figuratively snuffed out.

However, only innovators, by perceiving the need for new information, rules, and goals, communicating about them, and trying them out, can make the changes that transform systems. As Margaret Mead said,

"Never doubt that a small group of thoughtful, committed citizens can change the world. Indeed, it's the only thing that ever has."

We have learned the hard way that it is difficult to live a life of material moderation within a system that expects, exhorts, and rewards consumption. But one can move a long way in the direction of moderation. It is not easy to use energy efficiently in an economy that produces energy-inefficient products. But one can search out, or if necessary invent, more efficient ways of doing things, and in the process make those ways more accessible to others.

Above all, it is difficult to put forth new information in a system that is structured to hear only old information. Just try, sometime, to question in public the value of more growth, or even to make a distinction between growth and development, and you will see what we mean. It takes courage and clarity to challenge an established system. But it can be done.

In our own search for ways to encourage the peaceful restructuring of a system that naturally resists its own transformation, we have tried many tools. The obvious ones are displayed in Meadows et al. (2004)—rational analysis, data, systems thinking, computer modelling, and the clearest words we can find. Those are tools that anyone trained in science and economics would automatically grasp. Like recycling, they are useful and necessary, and they are not enough.

We don't know what will be enough. But we would like to conclude by mentioning five other tools we have found *helpful*. We introduced and discussed this list for the first time in our 1992 book. Our experience since then has affirmed that these five tools are not optional; they are essential characteristics for any society that hopes to survive over the long term. We present them here again in our concluding chapter "not as *the* ways to work toward sustainability, but as *some* ways." (Meadows et al. 1992)

"We are a bit hesitant to discuss them," we said in 1992, because we are not experts in their use and because they require the use of words that do not come easily from the mouths or word processors of scientists. They are considered too 'unscientific' to be taken seriously in the cynical public arena."

What are the tools we approached so cautiously? They are: visioning, networking, truth-telling, learning, and loving.

It seems like a feeble list, given the enormity of the changes required. But each of these exists within a web of positive loops. Thus their persistent and consistent application initially by a relatively small group of people would have the potential to produce enormous change—even to challenge the present system, perhaps helping to produce a revolution.

"The transition to a sustainable society might be helped," we said in 1992, "by the simple use of words like these more often, with sincerity and without apology, in the information streams of the world." But we used them with apology ourselves, knowing how most people would receive them.

Many of us feel uneasy about relying on such 'soft' tools when the future of our civilization is at stake, particularly since we do not know how to summon them up, in ourselves or in others. So we dismiss them and turn the conversation to recycling or emission trading or wildlife preserves or some other necessary but insufficient part of the Sustainability Revolution —but at least a part we know how to handle.

So let's talk about the tools we don't yet know how to use, because humanity must quickly master them.

Visioning

Visioning means imagining, at first generally and then with increasing specificity, what you really want. That is, *what you really want,* not what someone has taught you to want, and not what you have learned to be willing to settle for. Visioning means taking off the constraints of 'feasibility' of disbelief and past disappointments, and letting your mind dwell upon its most noble, uplifting, treasured dreams.

Some people, especially young people, engage in visioning with enthusiasm and ease. Some find the exercise of visioning frightening or painful, because a glowing picture of what *could be* makes what *is* all the more intolerable. Some people never admit their visions, for fear of being thought impractical or 'unrealistic'. They would find this paragraph uncomfortable to read, if they were willing to read it at all. And some people have been so crushed by their experience that they can only explain why any vision is impossible. That's fine; skeptics are needed too. Vision needs to be disciplined by skepticism.

We should say immediately, for the sake of the skeptics, that we do not believe vision makes anything happen. Vision without action is useless. But action without vision is directionless and feeble. Vision is absolutely necessary to guide and motivate. More than that, vision, when it is widely shared and firmly kept in sight, does *bring into being new systems.*

We mean that literally. Within the limits of space, time, materials, and energy, visionary human intentions can bring forth not only new information, new feedback loops, new behavior, new knowledge, and new technology, but also new institutions, new physical structures, and new powers within human beings. Ralph Waldo Emerson recognized this profound truth 150 years ago:

"Every nation and every man instantly surround themselves with a material apparatus which exactly corresponds to their moral state, or their state of thought. Observe how every truth and every error, each a thought of some man's mind, clothes itself with societies, houses, cities, language, ceremonies, newspapers. Observe the ideas of the present day ... see how each of these abstractions has embodied itself in an imposing apparatus in the community, and how timber, brick, lime, and stone have flown into convenient shape, obedient to the master idea reigning in the minds of many persons.... It follows, of course, that the least change in the man will

change his circumstances; the least enlargement of ideas, the least mitigation of his feelings in respect to other men ... would cause the most striking changes of external things.[4]

A sustainable world can never be fully realized until it is widely envisioned. The vision must be built up by many people before it is complete and compelling. As a way of encouraging others to join in the process, we'll list here some of what we see when we let ourselves imagine a sustainable society we would like to live in—as opposed to one we would be willing to settle for. This is by no means a definitive list. We include it here only to invite you to develop and enlarge it.

- Sustainability, efficiency, sufficiency, equity, beauty, and community as the highest social values.

- Material sufficiency and security for all. Therefore, by individual choice as well as communal norms, low birth rates and stable populations.

- Work that dignifies people instead of demeaning them. Some way of providing incentives for people to give their best to society and to be rewarded for doing so, while ensuring that everyone will be provided for sufficiently under any circumstances.

- Leaders who are honest, respectful, and more interested in doing their jobs than in keeping their jobs, more interested in serving society than in winning elections.

- An economy that is a means, not an end; one that serves the welfare of the environment, rather than vice versa.

- Efficient, renewable energy systems.

- Efficient, closed-loop materials systems.

- Technical design that reduces emissions and waste to a minimum, and social agreement not to produce emissions or waste that technology and nature can't handle.

- Regenerative agriculture that builds soils, uses natural mechanisms to restore nutrients and control pests, and produces abundant, uncontaminated food.

- Preservation of ecosystems in their variety and human cultures living in harmony with those ecosystems; therefore, high diversity of both nature and culture, and human appreciation for that diversity.

- Flexibility, innovation (social as well as technical), and intellectual challenge. A flourishing of science, a continuous enlargement of human knowledge.

- Greater understanding of whole systems as an essential part of each person's education.

- Decentralization of economic power, political influence, and scientific expertise.

- Political structures that permit a balance between short-term and long-term considerations, some way of exerting political pressure now on behalf of our grandchildren.

- High skills on the part of citizens and governments in the arts of non-violent conflict resolution.

- Media that reflect the world's diversity and at the same time unite cultures with relevant, accurate, timely, unbiased, and intelligent information, presented in its historic and whole-system context.

- Reasons for living and for thinking well of oneself that do not involve the accumulation of material things.

Networking

We could not do our work without networks. Most of the networks we belong to are informal. They have small budgets, if any, and few of them appear on rosters of world organizations[5]. They are almost invisible, but their effects are not negligible. Informal networks carry information in the same way as formal institutions do, and often more effectively. They are the natural home of new information, and out of them new system structures can evolve.[6]

Some of our networks are very local, some are international. Some are electronic, some involve people looking each other in the face every day. Whatever their form, they are made up of people who share a common interest in some aspect of life, who stay in touch and pass around data and tools and ideas and encouragement, who like and respect and support each other. One of the most important purposes of a network is simply to remind its members that they are not alone.

A network is non-hierarchical. It is a web of connections among equals, not held together by force, obligation, material incentive, or social contract, but by shared values and the understanding that some tasks can be accomplished together that could never be accomplished separately.

We know of networks of farmers who share organic pest control methods. There are networks of environmental journalists, 'green' architects, computer modellers, game designers, land trusts, consumer cooperatives. There are thousands and thousands of networks that developed as people with common purposes found each other. Some networks become so busy and essential that they evolve into formal organizations with offices and budgets, but most come and go as needed. The advent of the Worldwide Web certainly has facilitated and accelerated the formation and maintenance of networks.

Networks dedicated to sustainability at both the local and the global levels are especially needed to create a sustainable society that harmonizes with local ecosystems while keeping itself within global limits. About local networks we can say little here; our localities are different from yours. One role of local networks is to help reestablish the sense of community and relation to place that has been largely lost since the Industrial Revolution.

When it comes to global networks, we would like to make a plea that they be truly global. The means of participation in international information streams are as badly distributed as are the means of production. There are more telephones in Tokyo, it is said, than in all of Africa. That must be even more true of computers, fax machines, airline connections, and invitations to international meetings. But once more the wonder of human inventiveness seems to provide a surprising solution in the form of the Web and cheap access devices.

One could argue that Africa and other underrepresented parts of the world should attend first to their needs for many things other than computers and Web access. We disagree; the needs of the underprivileged cannot be effectively communicated, nor can the world benefit from their contributions, unless their voices can be heard. Some of the greatest gains in material and energy efficiency have come in the design of communications equipment. It is possible within a sustainable ecological footprint for everyone to have the opportunity for global as well as local networking. We must close the 'Digital Divide'.

If some part of the Sustainability Revolution interests you, you can find or form a network of others who share your particular interests. The network will help you discover where to go for information, what publications and tools are available, where to find administrative and financial support, and who can help with specific tasks. The right network will not only help you learn, but will allow you to pass your learning on to others.

Truth-Telling

We are no more certain of the truth than anyone is. But we often know an untruth when we hear one. Many untruths are deliberate, understood as such by both speaker and listeners. They are put forth to manipulate, lull, or entice, to postpone action, to justify self-serving action, to gain or preserve power, or to deny an uncomfortable reality.

Lies distort the information stream. A system cannot function well if its information streams are corrupted by lies. One of the most important tenets of systems theory, for reasons we hope we have made clear in our book *Limits to Growth—The 30-Year Update* is that information should not be distorted, delayed, or sequestered.

"All of humanity is in peril," said Buckminster Fuller, "if each one of us does not dare, now and henceforth, always to tell only the truth and all the truth, and to do so promptly—right now." (Buckminster Fuller 1981)

Whenever you speak to anyone, on the street, at work, to a crowd, and especially to a child, you can endeavor to counter a lie or affirm a truth. You can deny the idea that having more things makes one a better person. You can question the notion that more for the rich will help the poor. The more you can counter misinformation, the more manageable our society will become.

Here are some common biases and simplifications, verbal traps and popular untruths that we run into frequently in discussing limits to growth. We think they need to be pointed out and avoided, if there is ever to be clear thinking about the human economy and its relationship to a finite earth.

Not: A warning about the future is a prediction of doom.
But: A warning about the future is a recommendation to follow a different path.
Not: The environment is a luxury or a competing demand or a commodity that people will buy when they can afford it.
But: The environment is the source of all life and every economy. Opinion polls typically show that the public is willing to pay more for a healthy environment.
Not: Change is sacrifice, it should be avoided.
But: Change is challenge, and it is necessary.
Not: Stopping growth will lock the poor in their poverty.
But: It is the avarice and indifference of the rich, which locks the poor into poverty; the poor need new attitudes among the rich,

	then there will be growth specifically geared to serve their needs.

Not: Everyone should be brought up to the material level of the richest countries.

But: There is no possibility of raising material consumption levels for everyone to the levels now enjoyed by the rich. Everyone should have his or her fundamental material needs satisfied. Material needs beyond this level should be satisfied only if it is possible, for all, within a sustainable ecological footprint.

Not: All growth is good, without question, discrimination, or investigation.

And not: All growth is bad.

But: What is needed is not growth, but development. Insofar as development requires physical expansion, it should be equitable, affordable, and sustainable, with all real costs counted.

Not: Technology will solve all problems.

And not: Technology does nothing but cause problems.

But: We need to encourage technologies that will reduce the ecological footprint, increase efficiency, enhance resources, improve signals, and end poverty.

And: We must approach our problems as human beings and bring more to bear on them than just technology.

Not: The market system will automatically bring us the future we want.

But: We must decide for ourselves what future we want. Then we can use the market system, along with many other organizational devices, to achieve it.

Not: Industry is the cause of all problems, or the cure.

Nor: Government is the cause or the cure.

Nor: Environmentalists are the cause or the cure.

Nor: Any other group (economists come to mind) is the cause or the cure.

But: All people and institutions play their role within the large system structure. In a system that is structured for overshoot, all players deliberately or inadvertently contribute to that overshoot. In a system that is structured for sustainability, industries, governments, environmentalists, and most especially economists will play essential roles in contributing to sustainability.

Not: Unrelieved pessimism.

And not: Sappy optimism.

But: The resolve to tell the truth about both the successes and failures of the present and the potentials and obstacles in the future.

And

Above all: The courage to admit and bear the pain of the present, while keeping a steady eye on a vision of a better future.

Not: The World3 model, or any other model, is right or wrong.

But: All models, including the ones in our heads, are a little right, much too simple, and mostly wrong. How do we proceed in such a way as to test our models and learn where they are right and wrong? How do we speak to each other as fellow modellers with an appropriate mixture of skepticism and respect? How do we stop playing right/wrong games with each other and start designing right/wrong tests for our models against the real world?

That last challenge, sorting out and testing models, brings us to the topic of learning.

Learning

Visioning, networking, and truth-telling are useless if they do not inform action. There are many things to *do* to bring about a sustainable world. New farming methods have to be worked out. New businesses have to be started and old ones have to be redesigned to reduce their footprint. Land has to be restored, parks protected, energy systems transformed, international agreements reached. Laws have to be passed and others repealed. Children have to be taught and so do adults. Films have to be made, music played, books published, websites established, people counseled, groups led, subsidies removed, sustainability indicators developed, and prices corrected to portray full costs.

Each person will find his or her own best role in all this doing. We wouldn't presume to prescribe a specific role for anyone but ourselves. But we would make one suggestion: Whatever you do, do it humbly. Do it not as immutable policy, but as experiment. Use your action, whatever it is, to learn.

The depths of human ignorance are much more profound than most of us are willing to admit. Especially at a time when the global economy is coming together as a more integrated whole than it has ever been, when that economy is pressing against the limits of a wondrously complex planet, and when wholly new ways of thinking are called for. At this time, no one knows enough. No leader, no matter how authoritative he or she pretends to be, understands the situation. No policy should be imposed wholesale upon the whole world. If you can not afford to lose, do not gamble.

Learning means the willingness to go slowly, to try things out, and to collect information about the effects of actions, including the crucial but not always welcome information that the action is not working. One can't learn without making mistakes, telling the truth about them, and moving on. Learning means exploring a new path with vigor and courage, being open to other peoples' explorations of other paths, and being willing to switch paths if one is found that leads more directly to the goal.

The world's leaders have lost both the habit of learning and the freedom to learn. Somehow a political system has evolved in which the voters expect leaders to have all the answers, that assigns only a few people to be leaders, and that brings them down quickly if they suggest unpleasant remedies. This perverse system undermines the leadership capacity of the people and the learning capacity of the leaders.

It's time for us to do some truth-telling on this issue. The world's leaders do not know any better than anyone else how to bring about a sustainable society; most of them don't even know it's necessary to do so. A Sustainability Revolution requires each person to act as a learning leader at some level, from family to community to nation to world. And it requires each of us to support leaders by allowing them to admit uncertainty, conduct honest experiments, and acknowledge mistakes.

No one can be free to learn without patience and forgiveness. But in a condition of overshoot, there is not much time for patience and forgiveness. Finding the right balance between the apparent opposites of urgency and patience, accountability and forgiveness is a task that requires compassion, humility, clear-headedness, honesty and—that hardest of words, that seemingly scarcest of all resources—love.

Loving

One is not allowed in the industrial culture to speak about love, except in the most romantic and trivial sense of the word. Anyone who calls upon the capacity of people to practice brotherly and sisterly love, love of humanity as a whole, love of nature and of our nurturing planet, is more likely to be ridiculed than to be taken seriously. The deepest difference between optimists and pessimists is their position in the debate about whether human beings are able to operate collectively from a basis of love. In a society that systematically develops individualism, competitiveness, and short-term focus, the pessimists are in the vast majority.

Individualism and short-sightedness are the greatest problems of the current social system, we think, and the deepest cause of unsustainability. Love and compassion institutionalized in collective solutions is the better alternative. A culture that does not believe in, discuss, and develop these better human qualities suffers from a tragic limitation in its options. "How good a society does human nature permit?" asked psychologist Abraham Maslow. "How good a human nature does society permit?" (Maslow 1971)

The Sustainability Revolution will have to be, above all, a collective transformation that permits the best of human nature, rather than the worst, to be expressed and nurtured. Many people have recognized that necessity and that opportunity. For example, John Maynard Keynes wrote in 1932:

"The problem of want and poverty and the economic struggle between classes and nations is nothing but a frightful muddle, a transitory and unnecessary muddle. For the Western World already has the resource and the technique, if we could create the organization to use them, capable of reducing the Economic Problem, which now absorbs our moral and material energy, to a position of secondary importance...."

"[Thus the] day is not far off when the Economic Problem will take the back seat where it belongs, and ... the arena of the heart and head will be occupied ... by our real problems—the problems of life and of human relations, of creation and behaviour and religion."[7]

Aurelio Peccei, the great industrial leader who wrote constantly about problems of growth and limits, economics and environment, resources and governance, never failed to conclude that the answers to the world's problems begin with a "new humanism" (Peccei 1981):

"The humanism consonant with our epoch must replace and reverse principles and norms that we have heretofore regarded as untouchable, but that have become inapplicable, or discordant with our purpose; it must encourage the rise of new value systems to redress our inner balance, and of new spiritual, ethical, philosophical, social, political, aesthetic, and artistic motivations to fill the emptiness of our life; it must be capable of restoring within us ... love, friendship, understanding, solidarity, a spirit of sacrifice, conviviality; and it must make us understand that the more closely these qualities link us to other forms of life and to our brothers and sisters everywhere in the world, the more we shall gain."

It is not easy to practice love, friendship, generosity, understanding, or solidarity within a system whose rules, goals, and information streams are geared for lesser human qualities. But we try, and we urge you to try. Be patient with yourself and others, as you and they confront the difficulty of a changing world. Understand and empathize with inevitable resistance;

there is resistance, some clinging to the ways of unsustainability, within each of us. Seek out and trust in the best human instincts in yourself and in everyone. Listen to the cynicism around you and have compassion for those who believe in it, but don't believe it yourself.

Humanity cannot pass through the adventure of reducing the human footprint to a sustainable level, if that adventure is not undertaken in a spirit of global partnership. Collapse cannot be avoided if people do not learn to view themselves and others as part of one integrated global society. Both will require compassion, not only with the here and now, but with the far and future. Humanity must learn to love the idea of leaving future generations a living planet.

Is anything we have advocated in this book, from more resource efficiency to more compassion, really possible? Can the world actually ease down below the limits and avoid collapse? Can the human footprint be reduced in time? Is there enough vision, technology, freedom, community, responsibility, foresight, money, discipline, and love, on a global scale?

Of all the hypothetical questions we have posed in this book, these are the most unanswerable, though many people will pretend to answer them. Even we—your authors—differ among ourselves when tallying the odds for and against. The ritual cheerfulness of many uninformed people, especially world leaders, would say the questions are not even relevant; there are no meaningful limits. Many of the informed are infected with the deep cynicism that lies just under the ritual public cheerfulness. They would say that there are severe problems already, with worse ones ahead, and that there's not a chance of solving them.

Both of those answers are based, of course, on mental models. The truth of the matter is that *no one knows*.

We have said many times in our book *Limits to Growth—The 30-Year Update* that the world faces not a preordained future, but a choice. The choice is between different mental models, which lead logically to different scenarios. One mental model says that this world for all practical purposes has no limits. Choosing that mental model will encourage extractive business as usual and take the human economy even further beyond the limits, to collapse.

Another mental model says that the limits are real and close, and that there is not enough time, and that people cannot be moderate or responsible or compassionate. At least not in time. That model is self-fulfilling. If the world's people choose to believe it, they will be proven right. The result will be collapse.

A third mental model says that the limits are real and close and in some cases below our current levels of throughput. But there is just enough time with no time to waste. There is just enough energy, enough material, enough money, enough environmental resilience, and enough human virtue to bring about a planned reduction in the ecological footprint of mankind: a Sustainability Revolution to a much better world for the vast majority.

That third scenario might very well be wrong. But the evidence we have seen, from world data to global computer models, suggests that it could conceivably be made right. There is no way of knowing for sure, other than to try it.

Endnotes

[1] This text was adapted from chapter 8 of *Limits to Growth: The 30-Year Update* by Meadows for inclusion in this book. The principal author of chapter 8 was Donella Meadows. Any references to "we" refer to the co-*authors of Limits* Donella Meadows, Jørgen Randers, and Dennis Meadows.

[2] See Saouma (1993); Saouma was director general of the UN FAO from 1976 to 1993.

[3] See Ruckelshaus (1989); Ruckelshaus was twice administrator of the U.S. Environmental Protection Agency in the 1970s and 1980s.

[4] Ralph Waldo Emerson, Lecture on War, delivered in Boston, March 1838. Reprinted in *Emerson's Complete Works* (1887) Houghton, Mifflin & Co, Boston, volume XI, p. 177.

[5] Examples of such informal networks known to the authors and in their field of interest are The Balaton Group, Northeast Organic Farming Association NOFA, Center for a New American Dream CNAD (www.newdream.org), Greenlist, Greenclips (www.greenclips.com), Northern Forest Alliance (www.northernforestalliance.org), Land Trust Alliance (www.lta.org), International Simulation and Gaming Association ISAGA (www.isaga.info).

[6] Such an intermediate step is illustrated by the ICLEI, an international association of (currently 450) local governments implementing sustainable development. See www.iclei.org.

[7] J.M. Keynes, foreword to *Essays in Persuasion (1932)* Harcourt Brace and Company, New York.

References

Buckminster Fuller R (1981) *Critical Path*. St. Martin's Press, New York

Emerson, RW (1838) *Lecture on War,* delivered in Boston. Reprinted in *Emerson's Complete Works* (1887) Houghton, Mifflin & Co, Boston, Volume XI, p. 177

Keynes JM (1932) *Essays in Persuasion.* Harcourt Brace and Company, New York

Maslow A (1971) *The Farthest Reaches of Human Nature.* Viking Press, New York

Meadows DH, Meadows DL and Randers J (1992) *Beyond the Limits.* Chelsea
    Green Publishing, White River Junction VT
Meadows DH, Randers J and Meadows DL (2004) *Limits to Growth: The 30-Year
    Update.* Chelsea Green Publishing, White River Junction VT
Peccei A (1981) *One Hundred Pages for the Future.* Pergamon Press, New York,
    pp. 184–85
Ruckelshaus WD (1989) Toward a Sustainable World. *Scientific American*, Sep-
    tember 1989, p. 167
Saouma E (1993) *Statement by the Director-General.* Report of the Conference of
    FAO, 27th Session, November 1993, Rome
Worster D ed (1988) *The Ends of the Earth.* Cambridge University Press, Cam-
    bridge, pp.11–12

'Factor Four' and Sustainable Development in the Age of Globalization

ERNST ULRICH VON WEIZSÄCKER

From Limits to Growth to Sustainable Development

The kickoff for global environmental concerns has surely been the publication in 1972 of the *Limits to Growth Report* to the Club of Rome. In brief, it said that there are natural limits to be observed and that humanity is doomed if it ignores these limits to economic growth. The Limits to Growth Report was indeed a milestone in the international debate on the future of mankind. But it has become remarkably quiet around the report since the early eighties, and one keeps wondering why. The main authors, the Meadows and Jørgen Randers wrote another book 20 years later (Meadows et al. 1992) where they deplore the neglect of their theme and essentially say that the situation has worsened more rapidly than they had predicted.

The reason for the public neglect lies in the fact that the formula and some of the assumptions made in the Limits Report were faulty. As a scientist I was not the least surprised that well-founded critique was made and would never discard the Limits to Growth Report on these grounds as wrong in its core thrust. However, those who do not like the thrust at all are quick to jump on those few methodological deficiencies. Let me mention three of them:

- **Resource depletion:** Some optimistic economists say that the 'reach' of depletable natural resources has always and for systematic reasons been in the vicinity of thirty years, because this is about the time span worth looking at for companies and states concerned with resource availability. They always look at the low hanging fruits and simply do not go to the trouble and costs of developing access to the higher hanging ones. Hence, such economists have considered the Limits Report's assump-

179

M. Keiner (ed.), The Future of Sustainability, 179–192.
© 2006 *Springer. Printed in the Netherlands.*

tion of a thirty years reach of gas and oil reserves a trivial statement on which no depletion forecasts should be built.

– **Pollution:** In 1972, pollution was the most visible environmental threat. A constant mathematical relation was established in the Limits Report between pollution and industrial output. This rigidity was wrong, for the simple reason that affluent societies were able to answer the challenge with pollution control technologies. Pollution control at the end of the pipe was actually very convenient for the business camp. It could always argue that only prospering companies could do the anti-pollution job properly and therefore had to be treated well by the state. The whole game of pollution control ended up in the 'inverted U curve' paradigm: Societies start poor and clean. In the process of industrialization they become rich and dirty. When they are rich enough to combat pollution, they finally become rich and clean (Figure 1).

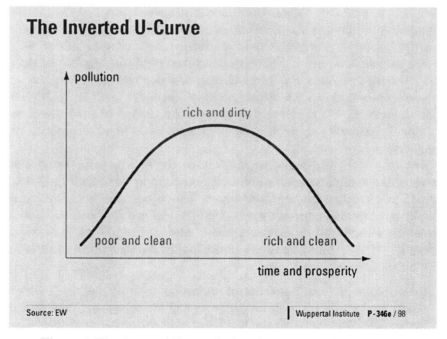

Figure 1. The 'inverted U-curve' of environmental pollution

This looks like the perfect, harmonious world. It is actually the basis of a whole new type of literature, that of the 'environmental optimists', for which the Danish statistician Björn Lomborg has perhaps become best known worldwide, in stark contrast to the pessimistic tone of the Limits Report.

– The static relations assumed between the five main factors used in the 'World 3 Scenarios' of the Limits to Growth Report: population, food, resources, industrial output and pollution control. In reality, technological progress can be said to consist in the decoupling of such parameters. As prosperity grew, the dynamics of population growth came to a dramatic halt, as can be seen in countries like Japan, Italy, Spain and certain Latin American countries, not to speak of northern Europe where this had been observed in 1972 already. Also the food industry nexus has always been subject to change, so far only in the favorable sense.

If the Limits to Growth Report is based on faulty assumptions, why am I still defending it? The reason is plain and simple. Because the main thrust of the Report is not really touched by these findings. Our world is and remains a limited one. Human societies must not and cannot grow beyond their natural limits. Some of the limits, however, lie not in resources or local pollution but in global environmental challenges, notably the greenhouse effect and the loss of biodiversity.

The new language, introduced by the Brundtland Report of 1987, refers to *sustainable development*. At the Earth Summit 1992 in Rio de Janeiro, the Agenda 21 was adopted to outline the immense tasks for sustainable development. Also, two major environmental conventions were ratified, the Climate Framework Convention and the Biodiversity Convention.

What these two conventions acknowledged is that our vision of 'rich and clean' may actually be quite *unsustainable* and the harmonious 'inverted U-curve paradigm' in itself is *only a myth*. That is because 'rich and clean' in its present meaning involves per capita consumption levels of depletable resources easily five to twenty times the rate of the 'poor and clean' stage. These depletable resources also include the absorptive capacity of the atmosphere. If six or ten billion people become 'rich and clean', exhaustion could come soon.

Another way to look at the sustainability challenge is to use Mathis Wackernagel and William Rees's (1997) concept of 'ecological footprints' to estimate how much environment in terms of area is needed to satisfy energy and resource consumption. They show that Americans, Germans or Japanese have footprints some ten times larger than those of the Chinese or the Indians. Using this analogy, the USA, Germany and Japan are hopelessly overpopulated, because they require much more territory than they have at their disposal, while China and India are not.

Or, more accurately stated, not yet. The developing countries, assisted by international investors, are working very hard to develop, i.e., to emulate Western styles of industrialization and consumption and thus acquire Western size footprints. By 2020, also China and India will be hopelessly overpopulated in terms of their footprints. We would need three to four Earths to accommodate six to eight billion US size ecological footprints. That is a rather drastic way of demonstrating that our present Western life-styles are ecologically *unsustainable* and collide with the limits to growth. In addition, it should be mentioned that the footprint analogy is based on the highly *artificial* assumption that roughly one third of the footprint area refers to energy consumption produced from renewables, because the authors wanted to calculate *sustainable* footprints for an unlimited time-span. But the remaining two thirds of the footprint area are real, made up of depletable resources used for such activities as growing wheat or oranges or cotton, for transport and housing or for trade and industry.

Regaining sustainable development was at the core of the agenda of the Earth Summit in Rio. Its Agenda 21 can be seen as a blueprint for our homework during the 21st century. However, two specific ecological challenges, biodiversity losses and the greenhouse effect, were too formidable to be simply included as chapters in Agenda 21. They bring additional dimensions to the whole problematic of sustainability.

Biodiversity and Dematerialization

Among the most alarming effects of civilization and economic growth is the rapid *loss of biodiversity*. At present, we are losing some twenty, perhaps fifty plant and animal species every day. This is mostly due to the destruction of natural habitats, which have been the home to hundreds of thousands of biological species, some of them, rather inconspicuous but nevertheless important in the interlinking webs of ecosystems. Habitat destruction mostly results from land conversion for mining, agricultural use, forest monocultures, or settlements. Developing countries tend to export most of the products of their lands. This is how we in the industrialized countries are able to maintain total 'footprints' exceeding our own territories. We 'export' many of our footprints to the South.

One reason for the massive land conversion and habitat destruction, perhaps the most important reason, are the gigantic *flows of materials* induced by our modern consumer society. Each one of us in the North induces material flows (or 'ecological rucksacks') of some forty, up to eighty tons per year.

I cannot see any plausible strategy of protecting what remains of our planet's biodiversity without *drastically* reducing the material flows traveling through the human technosphere. However, the Biodiversity Convention adopted at Rio de Janeiro does not even mention the nexus to dematerialization.

I shall come back to this point after discussing if a massive reduction in material intensity is at all feasible. How much dematerialization do we need if we want to allow developing countries to reach our levels of prosperity while simultaneously reducing the pressures on land for wildlife and biodiversity? Rough estimates by my friend Friedrich Schmidt-Bleek (1994) suggest that we shall need at least a factor of ten to reduce the ecological rucksacks in the West. That is quite a civilizational challenge.

Climate Change and the Energy Dilemma

Materials is one part of the story. The other is energy. A sizeable amount of 'footprints' are actually not at the ground but blasted into the air: human-caused greenhouse gas emissions. We are significantly changing the chemical composition of the atmosphere.

The Intergovernmental Panel on Climate Change has come up with projections of temperature rises during this century of 1.4 up to 5.8°C (Figure 2).

By 2020, the carbon dioxide concentrations will have doubled from pre-industrial levels. Insurance companies, notably reinsurers, fear increasing numbers of storms and floods. Annual damages have already exceeded USD 50 billion. If the climate develops further as some climatologists foresee, the countries worst hit will be developing countries, not to mention small island states which in the worst case will literally be washed away.

The scientific basis for such fears lies in the famous correlation between CO_2 concentrations and temperatures, discovered by excavations from Antarctic ice of air bubbles up to 160,000 years old (Figure 3).

More alarming is the correlation between these two and a *third* parameter, the seawater table, which can vary by some two hundred meters (Figure 4).

The geography of the coastlines, therefore, has changed dramatically in geological times. Not all of the changes, however, are due to global temperatures but to changes in geological structures.

Theoretically, the flood can come in a matter of a few decades. According to Michael Tooley (1989), the better part of the ice masses over Labrador and the Hudson Bay were breaking off into the sea some 7,800 years ago, letting the global seawater table rise by some 7–8 meters (Figure 5).

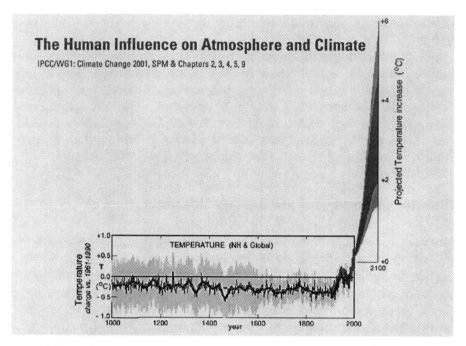

Figure 2. IPCC projections of global warming during the 21st century

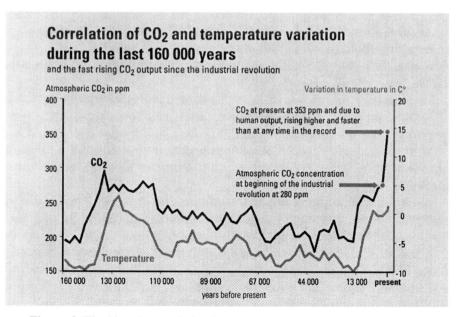

Figure 3. The historic correlation between CO_2 and temperatures on earth over the last 160,000 years

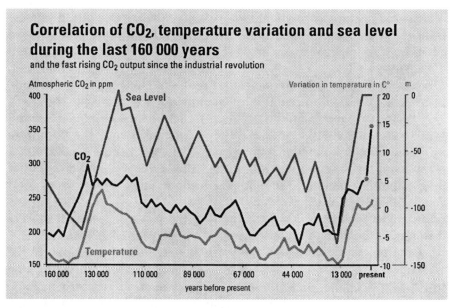

Figure 4. Also the sea water table is correlated with global temperatures

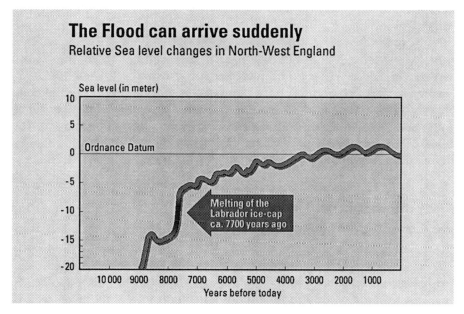

Figure 5. The flood can arrive suddenly

I am not suggesting that anything of this kind is *likely* to happen during the next fifty years. But we have no certainty that it will not happen. And we are increasing the probability of such disastrous events every year with global warming.

What can we do to stop the dangerous trends of climatic change? Climatologists recommend a reduction of greenhouse gas emissions by some 50–80 percent by the middle of our century. This would enable us to stabilize CO_2 concentrations at present levels. On the other hand, we learn from the World Energy Council that the demand for energy services and with it the emissions of carbon dioxide is going to rise steeply, most likely doubling within that period. That is at least a gap as large as a factor of four, which will have to be closed.

Some energy analysts say we can close the gap by turning to nuclear from fossils. But today, nuclear is a mere six percent of the world energy pie. Even this is subject to severe conflicts, and only a small part of the risks are covered by private insurance contracts. Imagine a neckbreaking rush towards tripling nuclear energy supplies in forty years—a political nightmare given the vulnerability of such installations to terrorism and war. What we would gain is an increase from six to eighteen percent of the pie. But because the pie itself is doubling, our gains drop back to a mere nine percent. This does not seem to be the key solution to the climate challenge.

With renewables, the substitution of fossils is a lot nicer but almost equally frustrating. Wind and solar make up 0.5% of the present pie. Let us assume an heroic strategy of increasing it *twentyfold*. Then we have reached ten percent of the present pie, but a mere five percent of the double sized pie. Hydro is used more at present, but please remember what nightmares are associated with present-day hydro schemes such as the famous Three Gorges Dam in China. We conclude this section by plainly stating that energy policy too is in a massive dilemma.

After the Industrial Revolution
the Eco-Efficiency Revolution

The challenges of sustainability, biodiversity protection and climatic change appear breathtaking. Lifestyles with unsustainably large footprints and yawning gaps of factors between four and ten in the areas of energy/climate change and material flows/biodiversity could leave us rather helpless. Fortunately, there is hope. Much of this hope is rooted in technological progress. But the task will be no smaller than the adventure of the Industrial Revolution.

What kind of animal is technological progress? All of us seem to assume that technological progress is an intangible 'natural' phenomenon that comes out of mix of scientific ingenuity and economic competition. States are said to have the best chance to accelerate it or impede it with bureaucracy or by setting unenlightened priorities. This standard picture of technological progress, I believe, is profoundly wrong. Technological progress runs in a direction that can be understood *and steered*.

In the past, technology was mostly driven (if not by military considerations) by the desire for economic expansion. The main emphasis was laid on the increase of *labor productivity*, which may have risen twentyfold during the last 150 years. That increased labor productivity becomes visible in the speed of our vehicles, in the power of our machines, in the organizational miracles of industrial production lines and in the unprecedented skills of modern information technologies.

The emphasis on labor productivity was very reasonable 150 years ago when human labor was extremely inefficient, and very strenuous as well. The winners in economic competition were almost always those who could offer more services and goods with less human labor. Wages rose more or less in proportion to the increase of labor productivity. As a result, workers were well-advised to support further productivity increases.

Nature's bounty appeared to be unlimited. So the exploitation of nature seemed like a legitimate and natural part of the game. Historians later called this game the Industrial Revolution. And it is still going on, worldwide.

Today, however, we are living in a completely different world from the early 19th century. Labor is now abundant, labor productivity is very high, and the real scarce resource is nature.

This means it is now high time to concentrate our efforts on increasing *resource productivity*. Even purely economic—and social—reasons speak for it. Slowing down the increase of labor productivity while speeding up resource productivity should make countries that have high levels of unemployment and import much of the natural resources they need, richer, not poorer.

Unemployment is a worldwide phenomenon. According to figures of the International Labor Organization (ILO), there are roughly 800 million jobless. This is a tragedy for their families and a disaster for the national budgets in countries where the state is obliged to pay unemployment benefits.

Shifting the emphasis to resource productivity should also be the best answer to the aforementioned challenge of sustainable development. The World Business Council for Sustainable Development often speaks of *eco-efficiency* as the new guiding term. And speaking on behalf of the Wuppertal Institute, I would like to add that efficiency should increase by a minimum of a factor of four. This is perhaps the only strategy allowing a reduction in size of the ecological footprints without jeopardizing employment and competitiveness. Since we are aiming at productivity jumps equally impressive as those characteristic of the Industrial Revolution, let us speak of the *Eco-Efficiency Revolution.*

The Good News: Factor Four

Before addressing the question of competitiveness, let us have a look at what is possible today in terms of dematerialization and energy productivity. Here I have some good news: *It is possible to quadruple resource productivity.*

Our book, *Factor Four,* coauthored with Amory Lovins, features fifty examples for increasing resource productivity by a factor of four at least (von Weizsäcker et al. 1997). Twenty examples were selected in the field of energy, twenty in material resource productivity and ten in transportation.

Let us have a look at the substance of the report. One very attractive example is what coauthor Amory Lovins has dubbed the *hypercar.* By almost entirely redesigning cars, making them lightweight and still crash-resistant and by using modern hybrid engines, the average fuel consumption can be pushed below 2 liters per 100km, which is more than four times better than today's fleets.

A few examples relate to the energy use of both private homes and office buildings. High tech insulation both of walls and of windows and an efficient heat exchange ventilation can reduce heating requirements by as much as a factor of ten. And applying modern prefabrication methods will help keep costs at very reasonable levels.

Other examples include lightbulbs, refrigerators, air conditioners, TV sets, mechanical fans, pumps and motors, computers and other office equipment. One is the success story of energy and waste savings over ten years at a big chemical firm in the USA with astounding average returns on investment far above 100%.

Renewable sources of energy will also play an important role in the efficiency revolution. They may not save energy by themselves. But they are at least 'carbon-efficient' and lend themselves to being combined with efficiency technologies, e.g., in the case of passive solar energy in buildings, optimized with the so-called translucent insulation technique.

Another very important sector of energy use is nutrition. By reducing the excessive use of fertilizers and the transportation of fodder, and by slightly cutting meat consumption, the energy requirements for a healthy diet can be cut by a factor of four.

The twenty examples of revolutionizing material productivity range from new construction methods and durable office furniture to water in homes to paper manufacturing and high tech recyclable plastics for wrapping and catering.

One striking feature of both energy and materials efficiency is how they strengthen the crafts sector of our economies. Insulating homes, using renewable energy, manufacturing and repairing durable goods are all crafts-based. Needless to say, modern craftsmen are well-equipped with modern communication technology. One example is tele-repair services, which allows an expert via a TV-connection to instruct customers in real time how to repair a broken dishwasher or other appliances. Another fine example is the replacement of a clumsy paper-based filing cabinet by a modern CD ROM system. There you save more than a factor of ten even if you generously include the 'ecological rucksacks' of the metal contained in the disks.

Similarly, the transport both of people and of goods can to a certain extent be replaced by information flows. Video conferences can—at least theoretically—save a lot of business travel. And e-mail needs much less resources than what has come to be known as 'snail mail'.

On the other hand, there are major rebound effects to be expected. For example, if you first 'meet' your business partners on the screen and your contacts are successful, you are more likely than before to want to see them in person. Thus, each video conference can be the cause of *additional* overseas travel.

Other examples from the transport sector relate to high tech measures increasing the capacity of existing railways and the cutting of ton kilometers for the production of strawberry yoghurt or fruit juices.

We calculated the effects of efficiency on the World 3 scenario. Leaving the mathematical relations as the Meadows team stated them but injecting two percent and four percent annual efficiency gains, we found an harmonious stabilization by 2100 or even by 2050, respectively. Thus, Factor Four can be considered a true answer to the challenges set out in the Limits to Growth Report.

Profitability, Long Term and Short Term

Needless to say, much of the efficiency revolution is not going to happen unless the frame conditions for doing business are changed. Efficiency must be made profitable.

To an astonishing extent, eco-efficiency is profitable now. Companies undergoing eco-audit procedures or even simply paying sufficient attention to the resource flows going through the firm have discovered that they gain considerable transparency also on the financial flows; they enjoy better cohesion with their staff and experience better customer relations. All this has led to the astonishing and most promising experience that portfolios of 'green' stocks can perform even better than the Morgan Stanley Capital International Index, which is seen by many as the benchmark index for shareholder value (Blumberg et al. 1997).

It is to be feared, however, that the potential for making profits by using eco-efficiency measures will be narrowly limited if present world market conditions prevail. These are characterized to a large degree by the widespread obsession with classical industrialization and the idea of local politicians that each and every industrial investment deserves their special attention and support. As a result, we see the most incredible amount of subsidies going into resource eating activities. As André de Moor (de Moor and Calamai 1997) of the Dutch Institute for Fiscal Studies has estimated, some 700 billion dollars are spent annually in the four fields of energy consumption, water, agriculture and motor transport. This does not yet account for all the tax advantages, free infrastructure and land given to investors. Desubsidizing resource use will be an important policy worldwide. But as in the case of pollution control, one country can hardly move if the competitors don't. Another, and related, policy tool is ecological tax reform. In a world of growing unemployment and of scarce natural resources, it just does not make sense to draw the biggest part of fiscal revenues from human labor while resource use goes essentially free of charge.

Let us have another look at the dynamics of neglect regarding resource efficiency. Tragically, the shareholder-value mentality has been pushed aggressively by US pension funds competing with each other on essentially nothing but financial yields. This has brought a mentality of short-termism into the financial world, which is tragic because the clients of pension funds can be expected to think long-term. They do not want spectacular quarterly reports but *sustainable* yields. Maybe a new front for legislation should be opened, obliging pension funds to some specified degree of long-termism. And states can give tax advantages to funds and individuals concentrating on eco-efficiency stocks.

Such new incentive systems directed towards long term profitability and sustainability will be honored by higher profits once our Western societies (and in their wake, others) begin to recognize that current practices are unsustainable. Realizing that a new technological revolution is waiting around the corner, enlightened states and business communities can initiate a rat race in that new direction. Then the eco-efficiency pioneers will harvest the fat dividends for a change. To initiate this process, international policy development will have to actively move towards new frame conditions conducive for the strategic increase of resource productivity.

Globalization

Ten years ago, I was even more optimistic about the paradigm shift towards green, high efficiency technologies. But then something happened, which has changed the world entirely, namely, 'globalization'. In my earlier role as Chairman of the Parliamentary Select Committee on Economic Globalization, I was surprised to learn that the very term *globalization* is actually quite new. It first began to play a role in the public realm in 1993.

The sudden appearance of the term globalization can be attributed to the end of the Cold War. Most of us were very happy that the Soviet empire collapsed, because we were suddenly freed from the spectre of a Third World War. But involuntarily, the Cold War protected the Western type of democracy, because it forced international capital always to seek consensus within society. That was gone after the collapse of the Soviet Union, and now we see the shareholder value mentality dominating the world.

This leads us to the *downside to globalization*. The increasing weakness of democracy in negotiating with the private sector was soon felt in the fields of taxation, environmental policy and social equity. In the OECD states, for example, there was a steady decrease of corporate tax rates, resulting from ever increasing pressures the private sector imposed on the states.

Unfortunately, there were also many losers, notably the poor in developing countries. Since the 1970's, the factor of the accumulated income of the richest 20 percent of the world population divided by the accumulated incomes of the poorest 20 percent rose from 30 to 75! In this situation, public priorities seem to be far away from the greenhouse effect and biodiversity losses.

After the failure of the WTO Ministerial in Cancún, I seem to hear more voices of people looking for a more inclusive kind of capitalism that does more justice between North and South. I should hope that this comes to pass and will also include more consideration for the environment and a program to redirect technological progress.

In the end, it is possible to have a rather optimistic outlook. The decoupling of well-being from resource use can happen rapidly and both private and public actors can play important roles in accelerating the transition. Unfortunately, I cannot prescribe the individual steps we need to take to make the future a more sustainable one, but I trust a spirit of innovation will do more than I possibly can.

References

Blumberg J, Blum G and Korsvold Å (1997) *Environmental Performance and Shareholder Value.* World Business Council for Sustainable Development, Geneva

de Moor A and Calamai P (1997*) Subsidizing Unsustainable Development; Undermining the Earth With Public Funds.* Earth Council, Toronto

Fussler C and James P (1997) *Driving Eco-Innovation.* Pitman, London

Meadows DH, Meadows D, Randers J and Behrens W (1972) *The Limits to Growth.* Universe Books, New York

Meadows DH, Meadows D and Randers J (1992) *Beyond the Limits.* McClelland and Stewart, Toronto

Schmidheiny S and the Business Council for Sustainable Development (1992) *Changing Course.* MIT Press, Cambridge MA

Schmidt-Bleek F (1994) *Wieviel Umwelt braucht der Mensch?* Birkhäuser, Basel

Tooley MJ (1989) Global Sea Levels: Floodwaters Mark Sudden Rise. *Nature* 342, pp. 20–21

Wackernagel M and Rees W (1997) *Unser ökologischer Fussabdruck.* Birkhäuser, Basel

von Weizsäcker E, Lovins A and Lovins H (1997) *Factor Four. Doubling Wealth, Halving Resource Use.* Earthscan, London

Ecological Footprint Accounting—
Comparing Earth's Biological Capacity
with an Economy's Resource Demand[1]

MATHIS WACKERNAGEL

Why Track Resource Consumption and Natural Capital?

Sustainability promises rewarding lives for all, now and in the future. Natural capital—nature's goods and services—is not the only ingredient in this vision. But without this type of capital—healthy food, energy for mobility and heat, fiber for paper, clothing and shelter, fresh air, and clean water—sustainability is impossible. This is why careful management of natural capital is central to current and future human well-being. Sustainability thus depends on protecting natural capital from systematic overuse; otherwise nature will no longer be able to provide society with these basic services.

How well are we using natural capital? Without measurements, we are blind and cannot effectively manage these essential natural resources. To take care of our natural capital, we must know how much we have and how much we use. This is no different from any financially responsible household, business, or government using accounts to keep track of its income and spending. Effective protection of our natural assets depends on accounts that keep track of humanity's demands on nature and nature's supply of ecological resources.

Ecological Footprint Accounts:
Capturing Human Demand on Nature

Ecological Footprint accounts are such balance sheets. They document for any given population the area of biologically productive land and sea

193

M. Keiner (ed.), The Future of Sustainability, 193–209.
© 2006 *Springer. Printed in the Netherlands.*

required to produce the renewable resources this population consumes, and to assimilate the waste it generates, using prevailing technology. This then can be compared to available areas. In other words, Ecological Footprints document the extent to which human economies stay within the regenerative capacity of the biosphere and who uses each portion of this capacity (Wackernagel and Rees 1996).

Such biophysical resource accounting is possible because resources and waste flows can be tracked, and because most of these flows can be associated with the biologically productive area required to maintain them. The Ecological Footprint area is expressed in global hectares—adjusted hectares that represent the average yield of all bioproductive areas on Earth. Since people use resources from all over the world and pollute far away places with their wastes, the Ecological Footprint accounts for these areas wherever they happen to be located on the planet.

Ecological Footprint Results

For each given year, Ecological Footprints compare human demand on nature with nature's regenerative capacity. Recent calculations by Global Footprint Network, published in WWF's *Living Planet Report 2004* (WWF et al. 2004), show that the average Swede required 7.0 global average hectares to provide for his or her consumption. If everyone on Earth consumed at this level, we would need four additional planets. The average Italian lived on a Footprint half that size (3.8 global hectares). The average Mexican occupies 2.5 global hectares, the average Indian lives on about one-third of that. The global average demand ('Footprint') is 2.2 global hectares per person (see Table 1).

Table 1. Comparison of the Ecological Footprint and the Biological Capacity of selected countries

	Population	Ecological Footprint	Biological Capacity	Ecological Deficit (-) or Reserve (+)[1]
	[millions]	[global ha/cap]	[global ha/cap]	[global ha/cap]
WORLD	6,148.1	2.2	1.8	−0.4
Argentina	37.5	2.6	6.7	4.2
Australia	19.4	7.7	19.2	11.5
Brazil	174.0	2.2	10.2	8.0
Canada	31.0	6.4	14.4	8.0
China	1,292.6	1.5	0.8	−0.8
Egypt	69.1	1.5	0.5	−1.0
France	59.6	5.8	3.1	−2.8
Germany	82.3	4.8	1.9	−2.9
India	1,033.4	0.8	0.4	−0.4
Indonesia	214.4	1.2	1.0	−0.2
Italy	57.5	3.8	1.1	−2.7
Japan	127.3	4.3	0.8	−3.6
Korea Republic	47.1	3.4	0.6	−2.8
Mexico	100.5	2.5	1.7	−0.8
Netherlands	16.0	4.7	0.8	−4.0
Pakistan	146.3	0.7	0.4	−0.3
Philippines	77.2	1.2	0.6	−0.6
Russia	144.9	4.4	6.9	2.6
Sweden	8.9	7.0	9.8	2.7
Thailand	61.6	1.6	1.0	−0.6
UK	59.1	5.4	1.5	−3.9
USA	288.0	9.5	4.9	−4.7
Combined	4,147.5	2.4	1.9	−0.5

Note that numbers may not always add up due to rounding. These Ecological Footprint results are based on 2001 data (WWF et al. 2004).

In contrast, the current supply ('biocapacity') of biologically productive land and sea on this planet adds up to 1.8 hectares per person. Less is available per person if we allocate some of this area to wild species, which also depend on it. Providing space for other species is necessary if we want to maintain the biodiversity that may be essential for the health and stability of the biosphere.

Comparing supply and demand, we see that in 2001 humanity's Ecological Footprint exceeded the Earth's biocapacity by over 20 percent (2.2 [gha/pers]/1.8 [gha/pers] = 1.2). In other words, it takes more than one

---

[1] In the last column, negative numbers indicate an ecological *deficit,* positive numbers an ecological *reserve.* All results are expressed in global hectares, hectares of biologically productive space with world-average productivity.

year and two months to regenerate the resources humanity consumed in that one year. Global demand began outpacing supply only recently, beginning in the 1980s. In 1961, for example, it took only 0.5 years to regenerate what was used in that year, as shown in the figure below (Wackernagel et al. 2002, WWF et al. 2004).

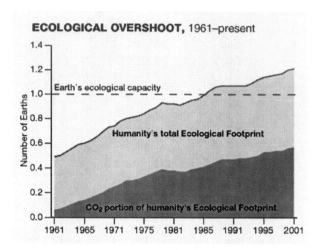

Figure 1. Human Demand versus number of
planets available

Human demand on the biosphere is increasing. Humanity's Ecological Footprint is shown here in number of planets, where one planet equals the total biologically productive capacity of the Earth in any one year. In 2001, humanity's Ecological Footprint was 2.5 times larger than in 1961, and exceeded the Earth's biological capacity by about 20 percent. This overshoot depletes the Earth's natural capital, and therefore is possible only for a limited period of time.

Overshoot and Ecological Deficit

Evidence is mounting that the sheer volume of resources flowing through the global economy is becoming today's key environmental challenge. Without counteraction, it has the potential to become tomorrow's key economic challenge.

The time delay between environmental impact and economic consequences stems from the fact that ecological limits—as financial budgets—can be overdrawn. It is possible to exceed global biocapacity because trees can be harvested faster than they regrow, fisheries can be depleted more rapidly than they restock, and CO_2 can be emitted into the atmosphere more quickly than ecosystems can sequester it. With humanity's current demand on nature, ecological deficit, or 'overshoot', is no longer merely a local but a global phenomenon. We are now consuming not only nature's interest, but also invading the principle.

Overshoot causes the liquidation of natural capital: carbon accumulates in the atmosphere, fisheries collapse, deforestation spreads, biodiversity is lost, and freshwater becomes scarce. Efficiency gains have helped to some extent: humanity's Ecological Footprint has grown slower than economic activities. Still, human demand on nature has steadily risen to a level where the human economy is now in global ecological overshoot.

Outlining the Accounting Method

The Research Question and Calculation Approach

The Ecological Footprint addresses on particular research question: *how much of the regenerative capacity of the biosphere is being occupied by human activities?* It does this, as declared above, by measuring how much biologically productive land and water area an individual, a city, a country, a region, or humanity uses to produce the resources it consumes and to absorb the waste it generates, using prevailing technology and resource management schemes. This land and water area can be located anywhere in the world.

Expressing use of biological natural capital in terms of area is useful, since life happens on surfaces. Primary producers—with help of photosynthesis—serve as the solar collectors for powering all animal species. Hence, surface areas matter, and most resource and waste flows can be measured in terms of the biologically productive area necessary to maintain these flows. (Those resource and waste flows that cannot be measured are excluded from the assessment. As a consequence, this assessment tends to underestimate the true Ecological Footprint).

Footprints can be analyzed from a consumption perspective[2] (as done in the national examples above), or at any stage of the production process. They can also be applied at all scales, from global down to any activity of organizations and populations, or for urban development projects, services, and products.

The Ecological Footprint uses a common, standardized measurement unit to make results comparable globally, similar to financial assessments that use one currency such as dollars or Euros to compare economies. The measurement units for Footprint accounts are global hectares. More precisely, a global hectare is 1 hectare of biologically productive space with world average productivity of the given year. When weighting each area in proportion to its usable resource productivity (that is, its annual production of usable resources and services), the different areas can be converted from hectares and expressed in a (different) number of global hectares of average productivity. 'Usable' refers to the portion of biomass used by humans, reflecting the anthropocentric assumptions of the Ecological Footprint measurement.

In 2001 (the most recent year for which consistent data is available)[3], the biosphere had 11.3 billion hectares of biologically productive area corresponding to roughly one quarter of the planet's surface. These 11.3 billion hectares include 2.3 billion hectares of water (ocean shelves and inland water) and 9.0 billion hectares of land. The land area is composed of 1.5 billion hectares of cropland, 3.5 billion hectares of grazing land, 3.9 billion hectares of forest land, and 0.2 billion hectares of built-up land.

Since these areas stand for mutually exclusive uses, and each global hectare represents the same amount of biomass production potential for a given year, they can be added up. This is the case for both the aggregate human demand (the Ecological Footprint) and the aggregate supply of biocapacity.

The Ecological Footprint calculated for each country includes the resources contained within the goods and services that are consumed by people living in that country, as well as the associated waste. Resources consumed for the production of goods and services that are exported to another country are added to the Footprint of the country where the goods and services are actually consumed, rather than of the country where they are produced.

The global Ecological Footprint is the area of productive biosphere required to maintain the material throughput of the human economy, under current management and production practices. Typically expressed in global hectares, the Ecological Footprint can also be measured in number of planets, whereby one planet represents the biological capacity of the Earth in a given year. Results could also be expressed, for example, in Austrian or Danish hectares, just as financial accounts can use different currencies.

The national analysis is based primarily on data published by the Food and Agriculture Organization of the United Nations (FAO), the International Energy Agency (IEA), UN Statistics Division (UNDESA, UN

Commodity Trade Statistics Database, UN Comtrade), and the Intergovernmental Panel on Climate Change (IPCC). Other data sources include studies in peer reviewed science journals or thematic collections.

These national accounts are now being maintained by the Global Footprint Network[4] and its partners. The purpose of this Network is to build global Footprint accounting standards with an 'open source' approach, so results become comparable and consistent across geography and time.[5]

Biocapacity and Bioproductivity

Biocapacity (biological capacity) is the total usable biological production capacity in a given year of a biologically productive area, for example within a country. It can be expressed in global hectares. Biologically productive area is land and sea area with significant photosynthetic activity and production of biomass. Marginal areas with patchy vegetation and non-productive areas are not included. There are 11.3 billion global hectares of biologically productive land and sea area on the planet. The remaining three-quarters of the Earth's surface, including deserts, ice caps, and deep oceans, support comparatively low levels of bioproductivity that are too dispersed to be harvested. Bioproductivity (biological productivity) is equal to the biological production per hectare per year. Biological productivity is typically measured in terms of annual biomass accumulation. Biocapacity available per person is calculated by dividing the 11.3 billion global hectares of biologically productive area by the number of people alive—6.15 billion in 2001—gives the average amount of biocapacity that exists on the planet per person: 1.8 global hectares.

What Footprint Accounts do NOT Include

The results of Footprint analyses tend to underestimate human demand on nature and overestimate the available biocapacity by

- choosing the more optimistic bioproductivity estimates when in doubt (e.g., carbon absorption);

- excluding human activities for which there are insufficient data (e.g., acid rain);

- excluding those activities that systematically erode nature's capacity to regenerate. They consist of:

– Using materials for which the biosphere has no apparent significant
  assimilative capacity (e.g., plutonium, heavy metals such as mercury
  and cadmium, and chlorinated hydrocarbons such as polychlorinated
  biphenyls (PCBs), dioxins, chlorofluorocarbons (CFCs); and
– Processes that irreversibly damage the biosphere, e.g., species extinction, fossil-aquifer depletion, deforestation, desertification.

If a hectare provides two or more services, that hectare is only counted
once in the Footprint. An example would be a hectare that grows timber
and collects drinking water. Counting the hectare only once in the Footprint makes sure there is no double counting, and allows hectares to be
added up.

The accounts include the productivity of cropland at the level of current
yields, with no deduction for possible degradation. However, if degradation takes place it will show up as reductions in future biocapacity assessments. The energy use for agriculture, including fertilizers, is included in
the energy Footprint.

Ecological Footprint calculations avoid double counting—that is, counting the same area twice. Consider bread: wheat is farmed, milled, and
baked, then finally eaten as bread. Economic data can track these sequen-
tial processes and report the amounts and financial values at each stage.
However, it is the same wheat grain throughout the production process that
finally is consumed by people. To avoid double counting, the wheat is
counted at only one stage of the process, while energy consumed at each
stage of the process is added to the Footprint.

The Carbon Footprint

The carbon Footprint from burning fossil fuel makes up about half of the
global Footprint. Since fossil fuel is not renewed at the rate of consumption, various methods are possible to capture how much of the biosphere's
regenerative capacity is occupied by this activity. Building on the UN
Framework Convention on Climate Change which aims at stabilizing the
CO_2 concentration in the atmosphere, the rationale is the following: Burning fossil fuel adds CO_2 to the atmosphere. The Footprint of fossil fuel is
calculated by estimating the biologically productive area needed to sequester enough CO_2 to avoid an increase in atmospheric CO_2 concentration.

Since the world's oceans absorb about 1.8 Giga tons of carbon every
year (IPCC 2001), only the remaining carbon emission is accounted for in
the Ecological Footprint. The current capacity of world average forests to
sequester carbon is based on FAO's Global Fiber Supply Model (FAO

2000) and corrected where better data are available from other FAO sources[6]. Sequestration capacity changes with both the maturity and composition of forests, and with shifts in bioproductivity due to higher atmospheric CO_2 levels and associated changes in temperature and water availability.

Other possible methods to account for fossil fuel use would result in even larger Footprints (Wackernagel and Monfreda 2004; Dukes 2003).

Applications of Ecological Footprint Accounts

The Ecological Footprint can be applied at scales ranging from single products to organizations, cities, regions, nations and humanity as a whole. It can be used to help budget limited natural capital. It also helps define the four complementary ways in which ecological deficits can be reduced or eliminated:

– Use resource-efficient technology that reduces the demand on natural capital;

– Reduce human consumption while preserving people's quality of life, for example reduce on the need for fossil fuels by making cities pedestrian friendly;

– Lower the size of the human family in equitable and humane ways so that total consumption decreases even if per capita demand remains unchanged; and,

– Invest in natural capital, for example by implementing resource extraction methods that increase rather than compromise the land's biological productivity, thereby increasing supply.

There have been Footprint applications on every continent. Global and national accounts have been reported in headlines worldwide, and over 100 cities or regions have assessed their Ecological Footprint (see some examples below). In California, Sonoma County's Footprint project *Time to Lighten Up* has inspired all cities of the county to sign up for the Climate Saver Initiative of the International Council for Local Environmental Initiatives (ICLEI). Wales has adopted the Ecological Footprint as its headline indicator. WWF International, one of the world's most influential conservation organizations, uses the Ecological Footprint in its communication and policy work for advancing conservation and sustainability. Government agencies, particularly in Europe, have studied the implication of Ecological Footprint results and have reexamined the significance of carrying capacity. A number of national ministers have repeatedly used the concept, including French President Jacques Chirac in his speech to the World Summit on Sustainable Development in Johannesburg.

"Our house is burning down and we're blind to it. Nature, mutilated and overexploited, can no longer regenerate and we refuse to admit it. Humanity is suffering. It is suffering from poor development, in both the North and the South, and we stand indifferent. The earth and humankind are in danger and we are all responsible.... We cannot say that we did not know! Let us make sure that the 21st century does not become, for future generations, the century of humanity's crime against life itself. This entails our collective responsibility. First and foremost the responsibility of the developed countries, who are frontrunners in terms of history, power and their consumption levels. If the whole of humanity were to behave like the Northern countries, it would take two more planets to satisfy our needs."

President Jacques Chirac (2002)[7]

Even larger media outlets are picking up the ideas: *The Economist* titled its July 2002 insert on the global environment "How many planets?" based on a Footprint assessment that showed it would take three planet Earths if all people lived OECD lifestyles.

Footprint Applications in Public Policy

Municipal Applications

– There may well be over one hundred Ecological Footprint studies for cities, ranging from student projects to comprehensive analyses of a metropolitan area's demand on nature. London, for instance, has already gone through three rounds. In 1995, urban sustainability expert Herbert Girardet estimated that the UK capital's Footprint was 125 times the size of the city itself. In other words, in order to function London required an area the size of the entire productive land surface of the UK to provide all the resources the city uses and to dispose of its pollutants and waste.

– In 2000, under the leadership of Mayor Ken Livingstone, *London* commissioned a more detailed Ecological Footprint study called City Limits. The report, sponsored by organizations including the Chartered Institution of Wastes Management, the Institution of Civil Engineers (ICE), and the Biffaward Programme on Sustainable Resource Use, was produced by Best Foot Forward and launched in September 2002. Results for this city and its 7 million inhabitants are available at: www. citylim-itslondon.com

- To respond to the challenges identified by the City Limits report, *London Remade,* a business membership organization supported by over 300 of the capital's major businesses and higher education institutions, wanted to analyze possible steps for reducing London's Footprint. In collaboration with London First, a waste management partnership, it commissioned consulting companies WSP Environmental and Natural Strategies to identify the reduction potential in a project called Toward Sustainable London: Reducing the Capital's Ecological Footprint. The first two of four reports, Determining London's Ecological Footprint and Priority Impact Areas for Action, are available at: www.londonremade.com

- Others have studied aspects of city living using the Ecological Footprint. For instance, the *Sustainable Consumption Group* of the Stockholm Environment Institute at York has led a number of studies of cities and regions (www.regionalsustainability.org/). They also contributed, with BioRegional, to a WWF-UK report called One Planet Living in the Thames Gateway, which identifies Footprint saving potentials for greener urban developments. The report is available at: www.wwf.org.uk/filelibrary/pdf/thamesgateway.pdf

- *Bill Dunster,* UK's leading ecological architect, uses the Footprint as the context for his designs. More on his work can be found at: www.zedfactory.com

National & Regional Applications
- A number of national and regional Footprint studies have contributed to policy discussions, some in close cooperation with government agencies. For example:

- *Wales* (pop. 2,900,000) The National Assembly for Wales adopted the Ecological Footprint as their headline indicator for sustainability in March of 2001, making Wales the first nation to do so. The first report was commissioned through WWF-Cymru and executed by Best Foot Forward. This report details Welsh energy, transportation and materials management. It can be found at: www.wwfuk.org/filelibrary/pdf/walesfootprint.pdf. An update of the report was produced by Stockholm Environment Institute and is available at: www.walesfootprint.org[8] (Barrett et al. 2005).

– *The State of Victoria, Australia* (pop. 4,650,000) EPA Victoria, the lead
  state agency responsible for protecting the environment, established a
  series of pilot projects in 2002 in partnership with a wide range of organizations and businesses to further investigate the practical applications of the Ecological Footprint to promote sustainability. See
  www.epa.vic.gov.au/eco-footprint.

– *Sonoma County, California* (30 miles north of San Francisco, pop.
  495,000) Under a grant from the U.S. EPA, Sustainable Sonoma
  County, a local NGO, used the Ecological Footprint as the foundation of
  a 2002 campaign. By inviting wide public participation and comment on
  the study before it was released, it was able to generate strong local buy-
  in. As a result, the launch of the study got countywide media coverage
  and built the groundwork for a subsequent campaign. The latter resulted
  in all municipalities of Sonoma County committing simultaneously to
  reduce their CO_2 emissions by 20 percent, making it the first U.S.
  county to do so. To meet this commitment, they established programs
  that track progress towards meeting their reduction goal. The Sonoma
  Footprint study is available at: www.sustainablesonoma.org/projects/
  scefootprint. html.

– *Six Southern regions of Italy.* Commissioned by WWF Italy, CRAS
  produced a study comparing the 6 southern regions of Italy. The study is
  available at: www.cras-srl.it/pubblicazioni/32.pdf

International Applications

– The *European Parliament* commissioned a comparative study on the
  application of Ecological Footprinting to sustainability. This study included case studies exploring potential uses of the Footprint in international legislation. The study, completed in 2001, was supervised by the
  Directorate General for Research, Division Industry, Research, Energy,
  Environment, and Scientific and Technological Options Assessment
  (STOA). It is available at www.europarl.eu.int/stoa/publi/pdf/00-09-
  03_en.pdf or as 10-page summaries in 11 European languages at:
  www.europarl.eu.int/stoa/publi/default_en.htm

– The *United Nations Population Fund* (UNFPA) report State of World
  Population 2001—Footprints and Milestones: Population and Environmental Change builds on Ecological Footprint concepts. See www. un-
  fpa.org/swp/2001/ english/ch03.html#5

An Indicator for 'Strong' and 'Weak' Sustainability

By monitoring human use of renewable natural capital, Ecological Footprint accounts provide guidance for sustainability: a Footprint smaller than the available biocapacity is a necessary condition for 'strong sustainability', a stance which asserts that securing people's well-being necessitates maintaining natural capital.

Some argue that 'strong sustainability' is too stringent since technology and knowledge can compensate for lost ecological assets. While this can be debated, even managing for 'weak sustainability' requires reliable accounting of assets. Hence, by measuring the overall supply of and human demand on regenerative capacity, the Ecological Footprint serves as an ideal tool for tracking progress, setting targets and driving policies for sustainability.

A Framework for Capturing the Ecological Consequences of Economic Choices: Shrink and Share

Sustainable Development is a commitment to improving people's well-being, while recognizing the existence of only one planet. Living within global limits requires from humanity to define these limits in realistic terms and find ways to allocate the 'maximum human demand' in ways acceptable to all nations. *Contraction & Convergence* as proposed by Aubrey Meyer from the Global Commons Institute (Meyer 2001) provides such a framework for globally allocating the right to emit carbon in a way that is consistent with the physical constraints of the biosphere. The approach rests on two transparent principles:

- **Contraction:** reducing humanity's emissions to a rate that the biosphere can absorb.

- **Convergence:** distributing total emissions in a way that is considered fair to all.

Although *Contraction & Convergence* focuses exclusively on CO_2 emissions, which are responsible for about 50 percent of humanity's Ecological Footprint, the *Contraction & Convergence* framework can be extended to other demands on the biosphere. We call this *Shrink & Share*. Shrinkage would occur when nations, organizations, and individuals reduce their Footprints so that consumption, production, investment, and trade activities do not exceed the regenerative capacity of the globe's life-supporting ecosystems. Sharing occurs if these reductions were allocated in ways considered fair by the participants.

This includes many possibilities: for example, it might imply that consumption, production, investment, and trade patterns change such that the per person Footprints in various nations deviate less and less from each other, that there is a more equitable distribution of the rights to use resources or that resource consumption rights are more closely tied to the resources a region or nation has available. Scenarios on how this might play out are published elsewhere.[9] But all scenarios compare ecological risk to effort to change the economy's path. In other words, it links choices about economic development with associated ecological risks.

Proactively reducing ecological deficits before being forced to do so is far preferable to the alternative, which can be considerably less pleasant. If planned for, deficit reductions brought about by decreasing demand for ecological resources need not necessarily entail hardship, and may even be associated with improvement in quality of life. On the other hand, history shows that when societies operating with an ecological deficit experience unplanned reductions in resource throughput and are forced to rely on their own biocapacity, a decline in quality of life, often severe, almost invariably follows (Diamond 2005).

What's in It for Governments and Regions?

Ecological Footprint accounts allow governments to track a city or region's demand on natural capital and to compare this demand with the amount of natural capital actually available. The accounts also give governments the ability to answer more specific questions about the distribution of these demands within their economy. For example, Footprint accounts reveal the ecological demand associated with residential consumption, the production of value-added products or the generation of exports; or they help assess the ecological capacity embodied in the imports upon which a region depends. This can help in understanding the region's constraints or future liabilities in comparison with other regions of the world, and in identifying opportunities to defend or improve the local quality of life.

Footprint accounts help governments become more specific about sustainability in a number of ways. The accounts provide a common language and a clearly defined methodology that can be used to support training of staff and to communicate about sustainability issues with other levels of government or with the public. Footprint accounts add value to existing data sets on production, trade and environmental performance by providing a comprehensive way to interpret them. For instance, the accounts can help guide 'environmental management systems' by offering a framework for gathering and organizing data, setting targets and tracking progress. The accounts can also serve environmental reporting requirements, and inform strategic decision making for regional economic development.

In addition, monitoring demand and supply of natural capital allows governments to

– Build a region's competitiveness by monitoring ecological deficits, since over time these deficits could become an increasing economic liability;

– Stay aligned with the business community's increasing focus on sustainability as a way to decrease future vulnerability;

– Manage common assets more effectively. Without an effective metric, these assets are typically valued at zero or less and their contribution to society is not systematically assessed nor included in strategic planning;

– Have access to an early warning device for long-term security that recognizes emerging scarcities and identifies global trends;

– Monitor the combined impact of ecological pressures that are more typically evaluated independently, such as climate change, fisheries collapse, loss of cropland, forestry overharvesting and urban sprawl;

– Identify local and global possibilities for climate change mitigation, and examine the tradeoffs between different approaches to atmospheric CO_2 reduction; and

– Test policy options for future viability and possible unintended consequences. For instance, it supports urban design processes, opens dialogue with stakeholders, helps manage expectations, provides a platform for sustainability management systems, supports training for sustainability, allows for ecological risk assessments, explain past successes more effectively.

Without regional resource accounting, countries can easily overlook or fail to realize the extent of these kinds of opportunities and threats. The Ecological Footprint, a comprehensive, science-based resource accounting system that compares people's use of nature with nature's ability to regenerate, helps eliminate this blind spot.

Endnotes

1   Written in collaboration with Dan Moran, Steven Goldfinger, and Josh Kearns.
2   Globally, the consumption Footprint equals the production Footprint. At the national scale, trade must be accounted for, so the consumption Footprint = production Footprint + imports − exports (assuming no significant change in stocks).
3   National accounts methodology build on Monfreda et al. (2004) an updated version of which can be downloaded from www.footprintnetwork.org. On this site, free academic licenses are available too, containing all the calculations. The Footprint is computed for all countries that are represented in UN statistical data back to 1961, with approximately 5,000 data points and 10,000 calculations per year and country. More than 200 resource categories are included, among them cereals, timber, fishmeal, and fibers. These resource uses are translated into global hectares by dividing the total amount consumed in each category by its global average productivity, or yield. Biomass yields, measured in dry weight, are taken from statistics (FAO 2004). Earlier methods were discussed in a special issue of Ecological Economics (2000).
4   Global Footprint Network, established as a non-profit organization in 2003, seeks to make the planet's ecological limits central to decision making by governments, businesses and households. It does this with its over 40 partner organizations from around the world by increasing the effectiveness and reach of the Ecological Footprint. Standardization of the accounting method is at the core of its strategy, with a first release of standards planned for early 2006. More on the science behind the Ecological Footprint and examples of how it has been used to advance sustainability can be found on the website www.footprintnetwork.org.
5   More about the standardization process and their progress is posted on the Network's website at www.footprintnetwork.org.
6   See FAO/UNECE (2000), FAO (1997), and FAO (2004).
7   www.un.org/events/wssd/statements/franceE.htm.
8   See also National Assembly for Wales (2004) Sustainable Development Indicators for Wales 2004. National Assembly for Wales, Statistical Bulletin 18/2004; www.wales.gov.uk/keypubstatisticsforwalesheadline/content/sustainable/2004/hdw20040323-e.htm.
9   The Living Planet Report: 2004 outlines four paths into the future and compares their risks. This builds on an earlier study of the *Shrink & Share* framework in the context of risk assessments and ecoinsurance schemes, as published in Lovink et al. (2004).

References

Barrett J, Birch R, Cherrett N and Wiedmann T (2005) *Reducing Wales' Ecological Footprint—Main Report,* March 2005. WWF Cymru, Cardiff, UK www.walesfootprint.org

Costanza R, Ayres R, Deutsch L, Jansson A, Troell M, Rönnbäck P, Folke C, Kautsky N, Herendeen R, Moffat I, Opschoor H, Rapport D, Rees W, Simmons C, Lewis K, Barrett J, Templet P, Van Kooten C, Bulte E, Wackernagel M and Silverstein J (2000) Commentary Forum: The Ecological Footprint. *Ecological Economics*, Vol. 32, No. 3 (2000), pp. 341–394

Diamond J (2005) *Collapse: How Societies Choose to Fail or Succeed*. Viking Penguin, New York

Dukes JS (2003) Burning Buried Sunshine—Human Consumption of Ancient Solar Energy. *Climatic Change*, Vol. 61 (1–2), pp. 31–44

FAO (1997) State of the World's Forests 1997. FAO, Rome, Italy.www.fao.org/ forestry/foris/webview/forestry2/index.jsp?siteId = 3321&sitetreeId = 10107 & langId = 1&geoId =, retrieved August 2004

FAO (2000) Global Fibre Supply Model. FAO, Rome, Italy. www.fao.org/forestry/ fop/fopw/GFSM/gfsmint-e.stm, retrieved August 2004

FAO (2004) FAOSTAT (FAO statistical databases). FAO, Rome, Italy. apps.fao.org, retrieved August 2004

FAO/UNECE (2000) *Temperate and Boreal Forest Resource Assessment 2000.* UNECE/FAO, Geneva, Switzerland

IPCC (2001) *Climate Change 2001: The Scientific Basis.* Cambridge University Press, Cambridge, UK

Lovink JS, Wackernagel M and Goldfinger SH (2004) *Eco-Insurance: Risk Management for the 21st Century.* Institute for Environmental Security, The Hague, Netherlands

Meyer A (2001) *Contraction and Convergence—The Global Solution to Climate Change.* Green Books, Dartington, UK

Monfreda C, Wackernagel M and Deumling D (2004) Establishing national natural capital accounts based on detailed ecological footprint and biological capacity accounts. *Land Use Policy* 21(3), pp. 231–246

UNDESA—United Nations Department of Economic and Social Affairs, Statistics Division. New York. unstats.un.org/unsd/ comtrade/, retrieved April 2005

Wackernagel M and Monfreda C (2004) Ecological Footprints and Energy. Cleveland CJ ed. *Encyclopedia of Energy*. Elsevier, Oxford

Wackernagel M and Rees WE (1996) *Our Ecological Footprint: Reducing Human Impact on the Earth.* New Society Publishers, Gabriola Island, British Columbia

Wackernagel M, Schulz NB, Deumling D, Linares AC, Jenkins M, Kapos V, Monfreda C, Loh J, Myers N, Norgaard R and Randers J (2002) Tracking the ecological overshoot of the human economy. *Proceedings of the National Academy of Sciences USA*, Vol. 99, No. 14, pp. 9266–9271

WWF—World-Wide Fund for Nature International, Global Footprint Network, UNEP World Conservation Monitoring Centre (2004) *Living Planet Report 2004*, WWF, Gland, Switzerland

Advancing Sustainable Development
and its Implementation Through Spatial Planning

MARCO KEINER

Current debate on sustainable development, stemming from its ambiguous universality (Campbell 2000) and the difficulties of its concrete implementation (Marcuse and Bartlett in this book) lead one to ask, "Are there better alternative means to achieving a more livable future?" Or, at the minimum, are there appropriate concepts to reemphasize the core aspect of sustainability: the responsible management of scarce resources?

In 1972, the Club of Rome released the report *Limits to Growth* (Meadows et al. 1972), relying on computer based dynamic systems modeling growth in human population and industrial production in a global system over the next century. Using selected resources-related indicators, simulations of the future, fed with different scenarios about the carrying capacity of the planet and availability of resources, demonstrated that humanity was about to destroy its living space. More than twenty years later, Meadows (1995) still insisted that sooner or later a scenario of collapse would be inevitable. For him, it is too optimistic to believe in sustainable development, for it is already too late to achieve this goal. As human behavior cannot be changed without obvious need (war, famine, etc.), the structures, pollution, and gaps that the present generation is about to leave to its heirs will inevitably force the next generations to strive for their sheer survival. In such a situation, individual and collective happiness is hardly achievable. Meadows calls this *Survivable Development*. Using this term, he comes back to the same nomenclature used by the 1972 UN Conference on the Human Environment at Stockholm (Friends of the Earth 1972). Is survivable development the only way into the future? Are there other approaches, which are more optimistic and try to change the main drivers of decline, overall awareness and behavior of the main users and abusers of planet Earth: humankind? In the following, I present an answer to these questions and outline how this answer might be implemented.

M. Keiner (ed.), The Future of Sustainability, 211–229.

Inter- and Intragenerational Equity and Welfare

Philosopher Vittorio Hösle,[1] following the thoughts of John Rawles, distinguishes three kinds of equity of distribution between humans: social equity, international equity, and equity between generations. The first two types comprise the problem of distribution between people living today and are widely recognized and thematized in current discourse. The third, 'equity between generations' or the fair relationship between present and future generations, is one that by nature is more difficult to grasp and the one I want to focus on.

As many other unsustainable societies throughout history, today's Western-lifestyle societies live in many domains on the capital of their children. Examples for the progressing destruction of the environment include ozone depletion, global warming, the disappearance of species, the deterioration of soils, overfishing of the oceans, the overharvesting of virgin forests, and atomic waste. The inappropriate and profit-oriented application of technological progress is responsible for long-term impacts of today's action (Lovelock 1979), which is based on the false belief that economic growth will, in time, lead to sustainable development (Daly 1993, Fritsch et al. 1994).

However, one might say that the contrary is the case. The effects of the construction of a nuclear power station, for example, last very far into the future because of the still unsolved problem of the ultimate disposal of atomic waste, which in turn will influence the quality of life of many generations. In addition, excessive national and communal debts have a negative impact on the ability of coming generations to act. These kinds of effects of contemporary policy offend against the principle of equity between generations: to leave a heritage that enables future generations to organize their life corresponding to their own visions and wishes and have at least the same potential opportunities at their disposal as current generations.

Pfister and Renn (1996) claim that the goal of sustainability oriented politics must be non-declining societal welfare levels over time. They define 'welfare' as the supply of all individual needs taking the distribution of resources into account. Thus, sustainability is a normative presetting on the distribution of resources between generations.

Philosophical Background for a New Approach

The early advocate of future ethics, philosopher Hans Jonas, coined a moral imperative, saying that human acting of today should leave enough freedom to future generations so that they will also be able to act. "Act so that the effects of your action are compatible with the permanence of genuine human life"; or simply, "Do not compromise the conditions for an indefinite continuation of humanity on earth"; or again with a positive spin, "In your present choices, include the future wholeness of Man among the objects of your will." (Jonas 1985) These reflections echo Kant's categorical imperative, "Act only on that maxim by which you can at the same time will that it should become a universal law." (Kant 1788).

Economist and philosopher Ralf Dahrendorf (1994) argues that 'life chances' (opportunities of living or potentials for decision) contrast to 'ligatures' (established bonds of the individual to society). Development offers new opportunities of choice and alternative action. The moral appeal that can be derived from this is to ensure that following generations will be afforded the precondition having at least as many options to act as we have.

In contrast, the pledge to prolong the present beatitude would lead to a neglect of the future, says philosopher Dieter Birnbacher (1999). The equity between generations is unbalanced when one generation is concerned primarily with short-term benefits. The short-term time horizon of the current generation is reflected in the Index of Sustainable Economic Welfare.[2] The ISEW, calculated for many countries, shows that while GDP is rising, sustainable economic welfare is falling. Our well-being today clearly threatens the well-being of coming generations. However, we merely push the costs for our bacchanal life into the future.[3] The situation is such that the happiness of the present-day adult is not only paid with the misery of yet unborn generations: the *futurization* of ecological problems already impacts the young generation of today. One example is the 'sustainable destruction' of habitats, a phenomenon that has broad reaching consequences on our living environment.

Considering the overwhelming evidence that current practices must be altered, as well as the idea of equity between generations, I proposed the *Principle of Good Heritage* (Keiner 2004).

The Principle of Good Heritage

Every generation inherits benefits and burdens from its previous generations. Similarly, each one, in pursuit of its well-being, shapes and transforms its living space and natural environment according to its needs. Today, for example, a lot of interstate highways exist. At first glance, this represents good heritage. But with this infrastructure comes the need to pay for its maintenance and the reinforced demand for automobiles. Among other pollutants, automobiles emit carbon dioxide, the substance responsible for the greenhouse effect. Because we have not yet found a solution to this problem and since most of the big global companies that could contribute to fundamental changes are profit- and not solution-oriented, our descendants will have to deal with the impact of these pollutants.

As the built environment and infrastructures have relatively long life-cycles, our descendants will also not be free to decide whether they would prefer this network of highways or if they would perhaps prefer another—environmentally and economically sounder—mode of traveling. While searching for alternatives in transportation, they probably will not only have to construct new transportation systems but will also have to dismantle existing structures at tremendous cost. Similar examples could be given for other inherited pollutions (water and air), for the organization of territory (settlement structures and functions), energy production, and so forth.

The *Principle of Good Heritage* is based upon the basic idea that we should strive to leave a *lesser burden* than that which we inherited ourselves. The task of today's generation should be to transform its heritage from burden to gain, from limitation to the freedom of acting, from struggling under a barely changeable fate to the potential of achieving happiness. The next generations should not only inherit living conditions that are *equal*, but even *better* than those we have today. Therefore, we will not only have to augment the social and economical but, first and foremost, the ecological values and qualities of life. In other words, our moral responsibility should be to increase the quality—and not only the quantity—of 'capital stocks' or ecological, human, and manmade resources (Serageldin and Steer 1994).

Karl-Raimund Popper wrote, "History has no sense," but he added that we can and must give it meaning, for example by affording the best *living chances* to as many people possible (Popper 1999).

Instead of just maintaining resource levels for those who will live on planet Earth when we are gone, our goal should be to explore and harness new resources and find substitutes for those that are non-renewable. The aim of our efforts should be to increase the quality, efficiency, and diversity of resources at our disposal. In so doing, not only inherited problems but also new opportunities could be offered to coming generations. The postulation is to act like farsighted testators who want their successors to be able to enjoy their heritage. However, this heritage won't come free. An effort from the heirs should be required. "That which you have inherited from your fathers, acquire it so as to make it your own",[4] demands the heirs to reflect about our ideas and values, successes and failures, and about their own chances and inconveniences. To prepare them, each generation should educate its children on sustainability issues and leave a kind of *testament* documenting its efforts, failures, recommendations, and hopes for future human life. Such a testament could be accompanied by a set of appropriate indicators (i.e., the Index of Sustainable Economic Welfare) with accompanying data that can later be used as evaluation and benchmarking tools.

The responsibility of both testators and heirs calls for the drafting of new *contracts between generations*. Such contracts would prevent the transfer of hereditary problems from our generation to the next, as well as ensure that positive parts of our heritage are handled with respect and care. In other words, the environmental, economic, social and institutional capital stocks should continue to grow in quality to improve the ecosystem and human well-being (Diener 1984). To work out such contracts between generations, *advocates of coming generations* should be appointed to *Future Councils* who will try to anticipate the expectations, values, and demands of those inheriting our world and act in their interests.

The Concept of Evolutionability

The augmentation of good heritage through the creation of new opportunities (life chances) and reduction of burdens is based on the theory that humanity continually evolves towards a higher quality of life. Until now, humanity could adapt to its living space or, expressed differently, adapt its living space to its needs. The negative impacts of human action, for example climate change, can be perceived today already and such signs will increase in the near future. At least in the foreseeable future, man-made climate change will remain as irreversible as the extinction of species, the destruction of natural habitats, and so forth. Humanity is destroying its living space. Waiting until humans will be able to settle on the planet Mars (Zubrin 1997) is not the right answer. Countermeasures have to be adopted before it is too late.

Instead of sustaining our burdens and limiting the freedom of our children and grandchildren, we should create an environment in which they do not have to be worried about survival but also be able to look ahead and reflect on new opportunities, developments and challenges. Thus, the Principle of Good Heritage leads to what I have tentatively defined as *evolutionable development* (Keiner 2004):

"Evolutionable development meets the needs of the present generation and enhances the ability of future generations to achieve well-being by doing so in a way that is as free as possible of inheritable burdens."

The vision of 'evolutionable development' encourages a society to neither waste nor destroy its means of existence. The use of raw materials (resources) and the stress on ecosystems should not go beyond the capacity of rehabilitation so that future generations will be able to live with the same or even more wealth than we do today. In this sense, evolutionable development is very close to what sustainable development would be, if the environment would clearly be the key element. Current models of sustainable development suggest the interconnectedness of environment, economy, and society. However, they do not accurately represent reality, which is an economy embedded in society and society embedded in the environment. Social and economical development can only take place if the environment offers the necessary resources: raw materials, space for new production sites and jobs, opportunities for quality-of-life (recreation, health etc.). Moreover, the environment provides a sink for waste. The ecosystem is therefore to be regarded as superior to the other dimensions of the triangle or prism models. These social, economical, and institutional aspects can only prosper if they adapt themselves to the limits of available environmental carrying capacity. To account for this, a change of attitude and of the economical, ecological and social behavior of the present generation is essential.

Furthermore, the above definition of 'evolutionable development' is more dynamic and action-oriented ('to enhance the ability') than the Brundtland definition of sustainability. Many definitions of sustainable development and their practical implementation reveal a more or less static character. 'To sustain' is often interpreted as 'to conserve' or 'to preserve' one or the other aspect of the status quo, giving an impression of enshrining or maintaining a certain state of development. However, sustainable development, in its proper sense, leads to continual qualitative improvement. In implementation efforts, the vital aspect of sustainability (as, for example, 'to develop', 'to promote', 'to improve', etc.) is often kept small. This minimizes the potential of required change and affords us the comfort of believing we are heading in the right direction when we are, in effect, standing still.

'Evolutionability' is not a new word. It is based on the term 'evolution' in its biological sense (Darwin, Lamarck) as well as in its philosophical sense (Spencer, Teilhard de Chardin, Bergson). 'Evolutionability' basically describes the capacity or tendency of a system to continually change to a higher or better state, in other words, the process of gradual and relatively peaceful social, political and economic advance. Today, computer scientists use the expression *evolutionability* in the field of network architecture to describe the inherent instability or tendency of a system to change. In addition, Tim Berners Lee, the inventor of the World Wide Web, uses a similar expression, *evolvability*, to describe the evolution of advanced computer languages (HTML and others) and the evolution of data on the web (Berners Lee 1998). In both cases, continual change and advancement are implied.

The concept of 'evolutionability' has much potential for providing the necessary impulse to readjust and strengthen efforts in sustainable development. In the following, I will argue that spatial planning on the regional level is a key field of policy activity for implementing sustainable development, and—based on this fact—the ideal realm to apply aspects of 'evolutionability' as well.

Implementing Sustainable Development Through Regional Spatial Planning

Spatial planning is the core discipline that steers the development of our present and future living space (the anthroposphere) and includes socioeconomic and environmental structures and their change, settlement area development, cultural imprints, the administrative organization of living space, and so forth. In many countries the implementation of sustainable development via spatial planning has been mandated. The central task of spatial planning is to organize the territory (of a country, region, city etc.) in a way that accommodates or mediates between multitudes of competing human uses (in the present as well as those projected in the future). Thus, spatial planning has to coordinate and ponder in an anticipatory way the space-relevant policies of all administrative levels and private stakeholders, harmonize covetousness, and offer visions for alternative forms of trend development. For this, appropriate planning methods, procedures, and instruments are required.

The question arises, on which political or administrative level sustainable and evolutionable development can be implemented best. In many countries, like the U.S., planning mainly takes place on the local level. Here, also sustainability oriented community initiatives as, for example, Local Agenda 21 processes are started. However, it is obvious that today's problems of spatial development cannot be solved on the local level alone. International and national levels, on the other hand, seem to be too abstract and too far away from where decisions are supported by the population. Therefore, the region appears to be the appropriate level where conceptual planning is to meet auto-determined implementation, backed by a strong identification with and knowledge of the unique characteristics of a living space.

Decentralization and Multi-Level Cooperation

The questions to be posed are *who* can determine what each region and its cities need to do in order to become sustainable and evolutionable, and *what* is a priority and what is not? This centers the discussion on the legitimacy of the decision-making process and its subjects. Strictly speaking, in a democratic society, the citizens and their representatives are clearly the ones who should be able to determine their needs and the direction of development. One main prerequisite for maintaining their fair share in decision-making is the proper vertical distribution of power between the national government and the regional and local levels.

As Hall and Pfeiffer (2000) state,

"Successful urban strategies will be possible only if national and local governments work in close cooperation, if central governments define more clearly the most efficient distribution of functions between the different levels of government, and if political activities follow a common framework."

Subsidiarity in the decision hierarchy is indispensable, i.e., subordinated authorities should be able to adapt and refine the general framework for their own specific territories. However, in practice, many relevant decisions are still made at the central government level in ministries that develop policies without a great deal of participation by regional and local players.

Although the importance of decentralization and the transfer of power and decision-making capability to the regional and municipal level for sustainable and evolutionable spatial development often uncontested, despite attempts to decentralize functions in many countries, the autonomy and power of regional and local governments continue to remain fairly modest. One consequence is that promising efforts in sustainable and evolutionable development remain sluggish and ineffective. The cause for this disconnect are outdated and weak administrative structures that continue to be accepted although they no longer serve the reality and needs of modern society.

Another point for consideration is the inefficient horizontal overlapping of functions and power of authorities:

"... Government structures need to work horizontally in order to implement the interdisciplinary nature of sustainable development. Governance is of core importance in implementing decisions towards sustainability and in effectively managing public interests in a politically organized community." (FARN, quoted in The Regional Environmental Center 2004)

If one accepts that the continual evolution of modern society is always changing the framework conditions of its environment, one has to be prepared to constantly reassess the appropriateness of the structures that control and manage it.

Orienting Spatial Development Plans Toward Sustainable Development

Today, spatial development in Switzerland, to take one example, cannot be judged to be sustainable (ARE 2005)—and thus, not evolutionable. A closer at Swiss cantonal structure plans reveals that almost all Swiss cantons have not yet formally oriented their spatial development toward the concept of sustainability. Still, an equitable development of all three pillars of sustainable development (environment, economy, and society) is not the central objective, but the creation of favorable conditions for economic growth. This is often formally described as "qualitative (economical) growth, considering environment and society" (Structure plan Canton of Berne, 2002). But only *considering* the environment, for example, is not the same as to plan *environmentally compatibly* or achieving equity among the three pillars of sustainability. Normally, planning decisions are still taken in favor of economic growth, accepting limited impacts on the pillars of environment and society. This is called 'weak sustainability', which, in the end, is 'no sustainability'.

Similar is the situation in Germany and Austria (Kanatschnig and Weber 1998) and certainly in many other countries worldwide. Therefrom stems the call for action to explicitly orientate spatial development toward sustainability to strengthen the effectiveness of existing planning instruments (Keiner et al. 2001, Healey 1993) as well as the enforcement of legally binding plans. As enforcement of planning measures is an important issue and evaluating the effectiveness of these policies in terms of achieving the stated goals is imperative. However, until today, there exist no 'ideal' planning instruments for achieving sustainability neither on the regional nor on the local level.

Visions and Planning Strategies
for Sustainable Spatial Development

The future task and main challenge for the management and planning of human living space will be to make the turnaround from rapidly growing urban areas worldwide, from social segregation and exclusion and the consumption of finite resources to sustainable and evolutionable development. In a first step, a greater emphasis should be placed on the clearer definition of what sustainable and evolutionable development is in the specific (historical, cultural, political, economical and environmental) context of each region concerned.

The public should be enabled and encouraged to actively work out long-term development visions for its living space. A model for such cooperation between planning authorities and the public could be the (local, regional) Agenda 21 approach. Here, individuals, associations and pressure groups can voluntarily work out a consensus on the future use of human living space and its resources. A number of authors underline the value of local communities and the importance of public participation (Douglass and Friedmann 1997, Malbert 1998, Holston 1999), a grassroots approach (Douglass 1995, Abers 2001) and the decentralization and democratization of planning decisions (Sandercock 1997 and 2002). FARN (quoted in The Regional Environmental Center 2004) also reinforces the importance of public participation:

"Public participation, different from the one existing in traditional representative democracies, must be present. Thus, a sustainable city requires institutions and systems that can facilitate public participation in decision-making regarding environmental use and management."

When planning and control overextend public authorities, self-help communities, NGOs, and grassroots initiatives become more and more important. These organizations contribute to problem-solving strategies on lower, basic levels, but often lack access to political power and capital. Authorities should support such organizations instead of arguing that they should follow 'official' policy instead.

To better screen the possible development alternatives, prospective and proactive scenarios can be used. Based on scenarios that meet broad-based consensus, visions or guidelines for future spatial development could be worked out. An important issue should be the needs of future generations and the acknowledgement of possible changes in life quality or increased burdens for them. To this end, *Future Councils* could be formed to facilitate this process, including *advocates for the needs of generations to come*. Once a common vision is coined and broad societal consensus is achieved, planning authorities will be able to derive the necessary strategies and measures for future regional and local development.

Making Sustainable Spatial Development Understandable and Acceptable

Sustainable Cities (1992) doubt that the environment-driven approach towards sustainability, as it appears in Europe and North America through *Local Agenda 21* processes, would work for developing countries because,

> *"It may be misleading to refer to many of the most pressing environmental problems in Third World cities as 'environmental' since they arise not from some particular shortage of an environmental resource but from economic or political factors which prevent poorer groups from obtaining them and from organizing to demand them."* (Sustainable Cities 1992)

Moreover, Rydin et al. (2003) doubt that sustainable development would be the most easily enacted response despite the obvious challenges of urbanization, because there lacks a consensus as to the meaning of sustainable development and quality of life, due to a real tension regarding future visions and what they entail. Also, Andam (2004) points out that there is a lack of vision and awareness in general, which has to do with illiteracy and language barriers in communication. Taking the example of Africa, "Environment and development conventions are conducted in languages that most local people cannot fully understand." (Andam 2004).

Therefore, it is indispensable that the concepts of sustainability and evolutionability be translated into simple language and understandable implementations on the grassroots level, based on a broad societal agreement that is fully assimilated into the culture of the organizations responsible for the implementation of spatial planning (Innes and Booher 2000).

Indicator-Based Monitoring and Controlling
of Sustainable Spatial Development

If sustainable and evolutionable development were to be better considered in local and regional planning, then the reorientation of the relevant planning objectives toward these principles would be a sine qua non. For this, a derivation of planning guidelines from overall principles for sustainability is necessary. Such requires the development of a coherent target system in order to narrow principles down to concrete objectives. In Switzerland, a guide for this task is offered in the form of a manual published by the Federal Office for Spatial Development ARE (INFRAS, ORL and C.E.A.T. 2001). In this guide, relevant sustainability targets, as well as indicators and target values are proposed for cantonal structure planning.[5]

However, spatial planning cannot become sustainable and evolutionable simply by following the relevant sustainability targets and their derived objectives. A regulation mechanism is needed that allows the steering of interventions (Priebs 1999, Szerenyi 1999). New operational specifications are necessary to detect deviations in the actual spatial development from the objectives, and to be able to apply corrective measures. Currently, in the context of an output oriented and cost effective *New Public Management* (NPM), first implemented in the 1990s in Anglo-Saxon countries, the instruments of 'monitoring' and 'controlling' are already in use in public administration. These instruments, which originate from the entrepreneurial marketing and management process, are being employed to increase the efficiency of public services. In the U.S., the concept of controlling has existed since the 1930s. In Western Europe enterprises, controlling was only adopted during the 1970s.

In the scientific community, the concept of controlling has already been widely used in cybernetic control loop models (Gaia hypotheses; Lovelock 1979). Controlling, in general, considers the goals defined by management and by processes designed to reach them. The constant comparison between the goal and the current actual state allows determining if the objectives are being achieved.

The hypothesis is that monitoring and controlling are also suitable for application in spatial planning (Keiner et al. 2001). In this context, these instruments would not be used to measure the performance of single administration units, but their plans as products of the Government that are binding for the entire public administration of the relevant territory. That means that focus is laid upon the planning document as central outcome of administrative work, its efficacy and processes, and not on the efficacy and processes inside the administration itself. With the easing of bureaucratic specifications, the decentralization of leadership responsibility and orientation of public service toward better performance in the process of NPM, there is also the tendency to displace control functions. Formerly, control applied to assessing the lawfulness of activities undertaken by the general public; today, control is more and more related to evaluating the output and effect of *administrative* initiatives.

Through continuous controlling, the working methodology of a plan may be changed, work methods may be simplified, and resource utilization may all be adjusted midstream. By adjusting goals, frameworks and indicators, new insights are gained (Hardi and Zdan 1997). Controlling is, in other words, "...the part of planning after you've decided what you wanted to be doing." (McNamara 1999) Although the term 'control' appeared in the 1947 British Planning Act, it was not meant to steer the whole process of implementing planning, but was limited to the control of development, i.e., the growth of settlements. It was based on the experiences of nineteenth century public health standards for new buildings. Development control was done through granting and refusing of planning permissions (Crow 1996, Booth 1999). This kind of control still exists today and is typically applied in Commonwealth countries (for example Australia, New Zealand, and Hong Kong) in the sense of avoiding urban sprawl and protecting landscape.

Essential for the monitoring and controlling of progress towards sustainable and evolutionable development is the proper choice of indicators (Prescott-Allen 1997, Bossel 1999). Indicators are also tools to focus public attention on that issue. They give an overview of whether a better quality of life can be achieved and a functional environment can be maintained, now and for generations to come. They can also be used to make comparisons over time and space to form the basis for spatial development policies. As Sustainable Measures (1998) put it,

"An indicator points to a problem or condition and helps to understand where we are, which way we go, and how far we are from where we want to be. Its purpose is to make a complex system comprehensible and perceptible, and to show how well these systems are functioning. A good indicator alerts us to a problem and shows where is need for action."

The use of indicators only makes sense if they are linked to measurable objectives and if supported by decision-makers. Planning needs "to anticipate future conditions—where we want to go, where we could go." (Hardi and Zdan 1997) In order to shape and evaluate spatial development it is necessary to define specific, quantifiable target values or a standardized measurement system for the objectives (Montgomery 1999). Defining realistic target values or goals is indispensable if sustainable development is to be achieved. A target value for evolutionable development could define, for example, the limits of costs for the maintenance and later rehabilitation of infrastructure projects for coming generations so that they will be able to financially bear the consequences of today's planning decisions. Whether a project is meeting its target values will be revealed by the changes in the monitored indicators.

In the UK, four planning regions (UK Office of the Deputy Prime Minister 1999) have worked out individual targets and indicators for Regional Planning Guidance (RPG). The Regional Development Agencies (RDA) will be subject to a monitoring and evaluation framework that is based upon five categories of indicators: (1) State of the region indicators, (2) RDA activity indicators, (3) strategic indicators, (4) program indicators, (5) efficiency indicators.

Other attempts with sustainability indicators on the regional level have already been made, for example, in New Zealand (Canterbury Regional Council) and Germany (North-Rhine-Westphalia). On the community level, mostly in the U.S. and the U.K., other examples exist like 'Sustainable Seattle', 'Santa Monica Sustainable City Program', 'Austin Sustainability Indicators Project', 'Sustainable Pittsburgh 2000', 'Coventry Sustainability Indicators', and so forth. On the national and supra-national levels, various projects and initiatives serve as a platform for learning from best practices. Examples, among others, include

- **European Common Indicators Initiative** ('Sustainable Cities Campaign' of the European Commission): In this initiative more than 100 cities that have signed the Charter of European Cities & Towns Towards Sustainability (The Aalborg Charter) in 1994 participate.

- **Global Urban Observatory**: A database of the UNCHS to measure living quality of more than 1,100 cities worldwide.

- **Urban Audit**: An initiative of the European Commission and Eurostat aiming at comparing the development of 58 big European cities.

In order to achieve not only sustainable but also evolutionable spatial development, specific indicators for monitoring and controlling spatial development plans and their impact have to be formulated. These indicators should be able to give answers to the following questions:

- Does the planning document take the needs of future generations into account? How? Is a contract between generations established? What is the content? Are there visions and strategies that go beyond our life span? Does it list measures to be taken in order to enhance the liberty of decision-making for future generations? Does it reserve surfaces for future use? Does it highlight the need for buildings that can be remanufactured (e.g., can office space, if no more needed, be transformed for living)? Does it define zones for solar settlements or other energy efficient types of settlement? Does it analyze the lifecycle of existing and future built environment; identify the future costs for renovation and rehabilitation?

– Has the planning document been worked out on a broad societal consensus and was an advocate for future generations present in the planning process? Were the persistence impacts of the planning projects analyzed? Were these projects designed to use regional renewable materials? If buildings and infrastructure have to be replaced, how much of the building material can be recycled?

– What will be the heritage for the next generation? In what sense is it better (or worse) than what we got from the previous generation? How many irreversible effects will the plan have?

– How do the energy and ecological balances change over time? Is the environmental condition getter better or worse? Which supplementary measures could be taken to mitigate problems?

Conclusion

If sustainable spatial development is to be achieved, new and innovative instruments are required to complete the existing planning tool palette. Much can be achieved, for example, in making indicators for controlling and monitoring development plans more effective and context-specific. To do so, basic questions regarding the steering of spatial development must be addressed involving the evaluating legal regulations and/or plans, interdictions, and/or agreements (e.g., Public-Private-Partnerships) influencing human behavior through charges and incentives.

In planning practice, pilot cities and regions for evolutionable development could be determined for the comprehensive testing of reforms and implementation tactics, and the guiding principles of spatial planning could be oriented towards the concept of evolutionability. Overall, planning instruments should be reshaped in order to create more environmental, economical, and socially just opportunities in the most efficient manner. Moreover, indicator based controlling of spatial development could allow planning authorities more flexibility in achieving sustainability targets, and enhanced public participation on the local and regional level would contribute to increasing overall awareness and sensibility for sustainability issues.

In conclusion, the concept of evolutionability is not meant to replace the principle of sustainability, but guide sustainable development in the desired direction: that the ability of future generations to meet their needs and to achieve collective and subjective well-being will not just be *not compromised*, but—expressed in positive terms—be improved.

"We don't know how the future will be. But we know that we have to act." (Friedrich Dürrenmatt)

Endnotes

[1] Quoted in SRzG—Stiftung für die Rechte zukünftiger Generationen (1999).
[2] ISEW; see Max-Neef (1995).
[3] Remember: *happiness* is, next to *life* and *liberty*, one of the key tenets of the U.S. declaration of independence.
[4] Goethe JW, Faust I
[5] The Swiss cantons correspond to the regional planning level in other states. However, there are also—sometimes very small—planning regions between the cantonal and municipal levels.

References

Abers R (2001) Practicing radical democracy—Lessons from Brazil. *DISP* 147, pp. 32–38

Andam K (2004) *Main Challenges of Sustainable Urban Development in Fast Growing Cities of the South.* Keynote speech. Annual Meeting of the Alliance for Global Sustainability, Gothenburg, March 23, 2004

ARE: Federal Office for Spatial Development (2005) *Raumentwicklungsbericht 2005,* Berne

Berners Lee T (1998) *Evolvability.* Keynote speech. 7th International WWW Conference. Brisbane, Australia, 15 April 1998

Birnbacher D (1999) Verantwortung für zukünftige Generationen—Reichweite und Grenzen. Mokrosch R and Regenbogen A eds. *Was heisst Gerechtigkeit? Ethische Perspektiven zur Erziehung, Politik und Religion,* pp. 62–81

Booth P (1999) From regulation to discretion—The evolution of development control in the British planning system 1909–1947. *Planning Perspectives,* Vol. 14, No. 3, pp. 277–289

Bossel H (1999) *Indicators for sustainable development—Theory, method, applications. A report to the Balaton group.* IISD International Institute for Sustainable Development, Winnipeg, Canada

Campbell H (2000) Sustainable Development—Can the vision be realized? *Planning Theory & Practice,* Vol. 1, No. 2, pp. 259–284

Crow S (1996) Development control—The child that grew up in the cold. *Planning Perspectives,* Vol. 11, No. 4, pp. 399–411

Dahrendorf R (1994) *Life chances.* University of Chicago Press, Chicago

Daly H (1993) Sustainable Growth—An impossible theorem. Daly H and Townsend K eds. *Valuing the Earth—Economics, Ecology, Ethics.* MIT Press, Boston, pp. 267–273

Dawkins R (1989) The evolution of evolvability. Langton CG ed. *Artificial Life,* Vol. 1. Addison-Wesley, Redwood CA, pp. 201–220

Diener E (1984) Subjective well-being. *Psychological Bulletin* 95, pp. 542–575

Douglass M (1995) *Urban Environmental Management at the Grass Roots. Toward a Theory of Community Activation,* East-West Center Working Papers, No. 42, Honolulu

Douglass M and Friedmann J eds (1997) *Cities for Citizens: Planning and the Rise of Civil Society in a Global Age.* John Wiley & Sons, Chichester, UK

Friends of the Earth (1972) *Declaration of the UN Conference on the Human Environment—Only one earth, an introduction to the politics of survival.* United Nations Conference on the Human Environment, Stockholm, 9–16 June 1972, Doc. A/Conference 48/14. London

Fritsch B, Schmidheiny S and Seifritz W (1994) *Towards an ecologically sustainable growth society—Physical foundations, economic transitions and political constraints.* Springer, Berlin, Heidelberg, New York

Hardi P and Zdan TJ (1997) *Assessing sustainable development—Principles in practice.* International Institute for Sustainable Development. Winnipeg

Hall P and Pfeiffer U (2000) *Urban Future 21—A Global Agenda for Twenty-First Century Cities.* Routledge, London

Healey P (1993) Planners, plans and sustainable development. Regional studies. *Journal of the Regional Studies Association.* Vol. 27, No. 8, pp. 769–776

Holston J (1999): *Cities and Citizenship.* Duke University Press, Durham

INFRAS, ORL and C.E.A.T. (2001) *Kantonale Richtplanung und nachhaltige Entwicklung—Eine Arbeitshilfe.* Bericht zuhanden des Bundesamtes für Raumentwicklung, Zurich

Innes J and Booher D (2000) Indicators for sustainable communities—A strategy building on complexity theory and distributed intelligence. *Planning Theory and Practice,* Vol. 1, No. 2, December, pp. 173–186

Jonas H (1985) *The Imperative of Responsibility—In Search of an Ethics for the Technological Age.* University of Chicago Press, Chicago

Kanatschnig D and Weber G (1998) *Nachhaltige Raumentwicklung in Österreich.* Schriftenreihe des Österreichischen Instituts für Nachhaltige Entwicklung, Band 4. Vienna

Kant I (1788) *Kritik der praktischen Vernunft.* Reclam, Leipzig

Keiner M (2004) Re-Emphasizing Sustainable Development—The Concept of 'Evolutionability': On living chances, equity, and good heritage. *Environment, Development and Sustainability,* Vol. 6, No. 4, pp. 379–392

Keiner M, Schultz B and Schmid WA (2001) Nachhaltige kantonale Richtplanung. *DISP* 146, pp. 18–24

Lovelock, JE (1979) *Gaia—A New Look at Life on Earth.* Oxford University Press

Malbert B (1998) *Urban Planning Participation: Linking Practice and Theory.* Dissertation, Department of Urban Planning and Design, School of Architecture, Chalmers University of Technology. Report SACTH 1998:1. Gothenburg

Max-Neef M (1995) Economic growth and quality of life. *Ecological Economics,* 15 (2), pp. 115–118

McNamara C (1999) *Management Function of Co-ordinating/Controlling—Overview of basic methods.* www.mapnp.org/library/cntrllng/cntrllng.htm, retrieved on March 12, 2004

Meadows DH (1995) It is too late to achieve sustainable development, now let us strive for survivable development. Murai S ed. *Towards Global Planning of Sustainable Use of the Earth—Development of Global Eco-engineering.* Elsevier, Amsterdam, pp. 359–374

Meadows DH, Meadows DL, Randers J and Behrens WW (1972) *The Limits to Growth.* Universe Books, New York

Montgomery A (1999) Best value and development control. *Planning* 1335, 10 September

Pfister G and Renn O (1996) *Ein Indikatorensystem zur Messung einer nachhaltigen Entwicklung in Baden-Württemberg.* Arbeitsbericht Nr. 64. Akademie für Technikfolgenabschätzung in Baden-Württemberg. Stuttgart

Popper KR (1999) *All Life is Problem Solving.* Routledge, London, New York

Prescott-Allen R (1997) *How do we know when we are sustainable?* IUCN Workshop 3c. Gland

Priebs A (1999) Neue Kooperationsstrategien zur Aufgabenerfüllung der Landes- und Regionalplanung. Akademie für Raumforschung und Landesplanung (ARL) ed. *Grundriss der Landes- und Regionalplanung.* Hannover, pp. 303–313

Rydin Y, Holman N, Hands V and Sommer F (2003) Incorporating Sustainable Development Concerns into an Urban Regeneration Project: How Politics can Defeat Procedures. *Journal of Environmental Planning and Management,* Vol. 46, No 4, pp 545–561

Sandercock L (1997) *Towards Cosmopolis: Planning for Multicultural Cities,* John Wiley & Sons, New York

Sandercock L (2003) Cosmopolis 2: *Mongrel Cities of the 21st Century.* Continuum Books, London

Serageldin I and Steer A eds (1994) *Making Development Sustainable: From Concepts to Action.* Environmentally Sustainable Development Occasional Paper Series No. 2. World Bank, Washington

SRzG: Stiftung für die Rechte zukünftiger Generationen (1999) *Was ist Generationengerechtigkeit?* Oberursel

Sustainable Cities (1992) Sustainable Cities: Meeting Needs, Reducing Resource Use and Recycling, Re-Use and Reclamation. *Environment and Urbanization.* Vol. 4, No. 2

Sustainable Measures (1998) *What is an indicator of sustainability?* www. sustainablemeasures.com/Indicators/WhatIs.html, retrieved on 25 July 2004

Szerenyi T (1999) *Indikatorensysteme nachhaltiger Regionalentwicklung auf unterschiedlichen räumlichen Massstabsebenen.* Working Paper No. 99-03 des Wirtschafts- und Sozialgeographischen Instituts der Universität zu Köln
The Regional Environmental Center (2004) *What is a Sustainable City?* Defined by FARN, Argentina www.rec.org/REC/Programs/SustainableCities/FarnArgentina.html, retrieved on 5 March 2004
UK Office of the Deputy Prime Minister (1999) *Scoping study RPG targets and indicators.* London
Zubrin R (1997) *The Case for Mars.* The Free Press, New York

Sustainability is Dead—Long Live Sustainability

ALAN ATKISSON

At the dawn of the Third Millennium, human civilization finds itself in a seeming paradox of gargantuan proportions. On the one hand, industrial and technological growth is destroying much of Nature, endangering ourselves, and threatening our descendants. On the other hand, we must accelerate our industrial and technological development, or the forces we have already unleashed will wreak even greater havoc on the world for generations to come.

We cannot go on, and we cannot stop. We must transform.

Facing a Great Paradox

At precisely the moment when humanity's science, technology, and economy have grown to the point that we can monitor and evaluate all the major systems that support life, all over the Earth, we have discovered that most of these systems are being systematically degraded and destroyed ... by our science, technology, and economy.

The evidence that we are beyond the limits to growth is by now overwhelming: the alarms include climatic change, disappearing biodiversity, falling human sperm counts, troubling slow-downs in food production after decades of rapid expansion, the beginning of serious international tensions over basic needs like water. Wild storms and floods and eerie changes in weather patterns are but a first visible harbinger of more serious trouble to come, trouble for which we are not adequately prepared.

Indeed, change of all kinds—in the Biosphere (nature as a whole), the Technosphere (the entirety of human manipulation of nature), and the Noösphere (the collective field of human consciousness)—is happening so rapidly that it exceeds our capacity to understand it, control it, or respond to it adequately in corrective ways. Humanity is simultaneously entranced

231

M. Keiner (ed.), The Future of Sustainability, 231–243.
© 2006 *Springer. Printed in the Netherlands.*

by its own power, overwhelmed by the problems created by progress, and continuing to steer itself over a cliff.

Our economies and technologies are changing certain basic structures of planetary life, such as the balance of carbon in the atmosphere, the amount of forest cover, genetic codes, species variety and distribution, and the foundations of cultural identity.

Unless we make technological advances of the highest order, many of the destructive changes we are causing to nature are irreversible. Extinct species cannot (yet) be brought back to life. No credible strategy for controlling or reducing carbon dioxide levels in the atmosphere has been put forward. We do not know how to fix what we're breaking.

At the same time, some of the very products of our technology—plutonium, for instance—require of us that we maintain a very high degree of cultural continuity, economic and political stability, and technological capacity and sophistication, far into the future. To ensure our safety and the safety of all forms of life, we must always be able to store, clean up, and contain poisons like plutonium and persistent organic toxins. Ultimately, we must be able to eliminate them safely. At all times, we must be able to contain the actions of evil or unethical elements in our societies who do not care about the consequences to life of unleashing our most dangerous creations. In the case of certain creations, like nuclear materials and some artificially constructed or genetically modified organisms, our secure custodianship must be maintained for thousands of years.

We are, in effect, committed to a high-technology future. Any slip in our mastery over the forces now under our command could doom our descendants—including not just human descendants, but also those wild species still remaining in the oceans and wilderness areas—to unspeakable suffering. We must continue down an intensely scientific and technological path, and we can never stop.

Sustaining such high levels of complex civilization and continuous development has never before happened in the history of humanity, so far as we know. From the evidence in hand, ancient civilizations have generally done no better than a few hundred years of highly variable progress and regress, at comparatively low levels of technology, with relatively minor risks to the greater whole associated with their inevitable collapse.

The only institutions that have demonstrated continuity over millennia are religions and spiritual traditions and institutions. So, while we must be intensely scientific, our future is also in need of a renewed sense of spirituality and the sacred. Given our diversity and historic circumstances, no one religion is likely to be able, now or in the future, to sustain us or unite us. We need a new sense of spirituality that is inclusive of believers, non-believers, and those for whom belief itself is not the core of spiritual experience. We need a sense of the sacred that is inclusive of the scientific quest and the technological imperative. We need a common sense of high purpose that connects, bridges, and uplifts all of our religious traditions to their highest levels of wisdom and compassion, while sustaining and honoring their unique historical gifts. We need, especially, all the inspiration and solace they can offer, because the task ahead of us is enormous beyond compare.

Our generation is charged with an unprecedented responsibility: to lay secure foundations for a global civilization that can last for thousands of years. To accomplish this task, we must, in the coming decades, maintain and greatly enhance our technical capacities and cultural stability, while simultaneously changing almost every technological system on which we now depend so that it causes no harm to people or the natural world, now or in the future.

Our situation is not only without precedent; it is virtually impossible to comprehend. Those who, in the waning decades of the Second Millennium, have been able to comprehend this Great Paradox to some degree often feel themselves emotionally overwhelmed and powerless to effect change—the situation I have elsewhere called 'Cassandra's Dilemma', after the mythical Trojan prophet whose accurate foresight went unheeded. Those in power, on the other hand, face stiff barriers to comprehension and action, including financial, political, and psychological disincentives. Denial and avoidance have been civilization's predominant responses to the warnings coming from science and the signals coming from nature during the 1970s, 80s, and 90s.

But the feedback from nature, as well as the growing global distress signals from those left behind in either relative or absolute poverty, are both becoming so strong that they can no longer be denied, even by those with the greatest vested interest in denial. These early decades of the Third Millennium—and especially this first decade, which philosopher Michael Zimmerman has said should be declared 'the Oughts' to signify the urgency for addressing what ought to be done—are the decades of reckoning, the time for decisively changing course.

Modest Changes are Not Enough

Change is clearly possible. Modest changes in the direction of greater sustainability are now underway, and modest, incremental changes in both technology and habitual practice can ameliorate—indeed, have ameliorated—some dangerous trends in the short run.

But overall, incremental change of this sort has proven exceedingly slow and difficult to effect, and most incremental change efforts fall far short of what is needed. Carbon emissions, which are now causing visible climate change, provide a good example: current global agreements for modest reductions are hard to reach, impossible to enforce, and virtually without effect; and even if they were successful, they would have a negligible impact on the critical trend. Far more dramatic changes are required.

Dramatic, rapid change, in the form of extremely accelerated innovation in the Noösphere (conscious awareness and understanding) and the Technosphere (physical practice) is necessary to prevent continuing and ever increasing catastrophic damage to the Biosphere as well as adapt to those irreversible changes to which the planet is already committed, such as some amount of climatic instability. The rapid evolution of the many social, economic, and political institutions, which mediate between the Noösphere and the Technosphere, is obviously necessary as well.

Without extraordinary and dramatic change, the most probable outcome of industrial civilization's current trajectory is convulsion and collapse. 'Collapse' refers not to a sudden or apocalyptic ending, but to a process of accelerating social, economic, and ecological decay over the course of a generation or two, punctuated by ever-worsening episodes of crisis. The results would likely be devastating, in both human and ecological terms. The onset of collapse is probably not ahead of us in time, but behind us: In some places, such as storm-ravaged Orissa, Honduras, Bangladesh, Venezuela, even England and France, collapse-related entropy may already be apparent.

Trend, of course, is probability, not destiny. It is still theoretically possible, albeit very unlikely, that civilization could continue straight ahead, without any conscious effort to direct technological development and the actions of markets in more environmentally benign and culturally constructive ways, and escape collapse through an unexpected (though currently unimaginable) technological breakthrough or improbable set of events. Some have called this the 'Miracle Scenario'.

But hoping for a miracle is by far the riskiest choice. The future may be fundamentally unknowable, but certain physical processes are predictable, given adequate knowledge about current trends, causal linkages, and systemic effects. Prediction based on extrapolation is not just the province of physics: much of our economy is focused on efforts to accurately predict the future based on past trends. The Internet economy, for example, relies upon Moore's Law (that the speed and capacity of semiconductor chips doubles roughly every 18 months). Insurance companies base their entire portfolio of investments and fees on statistical assessments of past disasters and projected trends into the future.

When it comes to the prospects for sustaining our civilization, we have to trust our species' best judgment, which comes from the interpretations and extrapolations of our best experts. These experts—such as the respected Intergovernmental Panel on Climate Change—are reporting a disturbingly high degree of consensus about the level of threat to our future well-being. We are in trouble.

We must transform our civilization.

Transformation is Possible

Dramatic civilizational change—transformation, in a word—is not so difficult to imagine. History is full of examples. Global history since the Renaissance, with all our remarkable transformations in technology, economics, and culture, is largely a product of humanity learning to take seriously the evidence of its senses, to reflect on that evidence carefully, and to make provisional conclusions that can be tested. This is the cornerstone of science.

If we are to take seriously the evidence of our senses and our science, we must provisionally conclude that we are now largely responsible for living conditions on this planet. We have the power to fundamentally shape climate, manage ecosystems, design life-forms, and much more. The fact that we are currently *doing* these things very badly obscures the fact that we are doing them, and can therefore learn to do them better. Designing and managing the world is now our responsibility. That is the hypothesis that must now be tested by humanity as a whole, if we are to prevent collapse and succeed in restoration.

To succeed, we must take our responsibility as world-shapers far more seriously than we currently do. History demonstrates that we, as a species, have the power to create the future we envision. If, therefore, we give in to despair, collapse will follow. If we cultivate a vision of ourselves as powerful and wise stewards of our planetary home, transformation becomes possible.

Examples of cultural transformation occurring within a generation or less abound. The Meiji Restoration transformed Japan from a closed, agricultural society to an industrial one in just a few decades. The wholesale redirection of the North American and European economies during World War II took just a few years. The Apollo Program's success in putting humans on the moon transpired, on schedule, within a decade. The fall of the Berlin Wall, the end of Apartheid, the change in China from a state-planned to a market economy ... much of recent history suggests that transformation is not only possible, but a frequent occurrence in civilizational evolution.

None of these events, however, remotely approaches the scale of global transformation we must now effect in technology, energy, transportation, agriculture, infrastructure, and economics, based on a new cultural understanding of our role as nature's managers, the world's architects, the planet's artists and engineers. But this testimony from history illustrates something profoundly important about transformation, in addition to its raw and indisputable possibility: no transformative change truly happens suddenly. Nor does transformation involve the magical or instantaneous creation of a new culture. 'Transformation' is the name we give to *the extremely accelerated adoption of existing innovations, together with the acceleration of innovation itself.*

Understanding transformation in these terms gives, to those who seek to create one, a reason for hope. An enormous amount of design work, preliminary to a transformation of the kind envisioned here, has already been done. Inventions, policies, models, scenarios, alternatives ... innovations of all kinds have been developed by thoughtful and committed people over a generation, and the speed of innovation is increasing. Intense and focused commitment by a critical mass of talented, dedicated, and influential people—in business, government, religion, the arts, the civil sector, every walk of life—could accelerate the process by which innovation enters the mainstream of technical and social practice, and thereby turns humanity on a more hopeful course.

By framing ambitious and visionary goals, and by highlighting the dangers and risks of inaction, this corps of skilled and forward-looking individuals in groups, organizations, corporations and governments could inspire others. The numbers involved could grow exponentially, and as institutions became thoroughly oriented toward achieving transformation, enormous resources could be mobilized, accelerating the transformation process still further.

One generation of intensely focused investment, research, and redevelopment—redesigning our energy systems, overhauling our chemical industries, rebuilding our cities, finding substitutes for wood and replanting lost forests, and so much more—could transform the world as we know it into something far more beautiful, satisfying, and sustainable.

This I believe: Sustainability is possible. Sustainability is desirable. Sustainability is a goal worthy of one's life's work. Sustainability is the great task of the next century. Sustainability is the next challenge on the road to our destiny.

Sustainability is Dead—Long Live Sustainability

The concept of 'sustainability' sprouted and spread like grass during the last few decades of the 20th Century. In scientific terms, it means a system state that can endure indefinitely. Consider a forest: by not losing trees any faster than they grow back, the forest 'system' survives despite (and sometimes because of) fires and other natural disturbances. The forest is sustainable. In more popular terms, 'sustainability' has come to mean long-term survival and well-being in general, both for human civilization and the rest of nature.

As a guide to the future, the word 'sustainability' is currently both our best hope and our biggest obstacle. Many have found the concept a great inspiration, and it has given rise to hundreds of initiatives around the world. But as a word, 'sustainability' tends to bore some people and frustrate others. Many have questioned the clarity of 'sustainability', and others have doubted its utility in practice. Indeed, the word is beset by problems; but the problems run deeper than most criticism would suggest.

As the new Millennium begins, sustainability, as a word, is dying. It is not, as some would claim, that there is too much vagueness in its definition. A process can either continue (sustainable), or it cannot (unsustainable). A society's use of its resources, its social patterns, and its pollution emissions are such that they will either go on indefinitely (sustainability), or they will not (collapse). Societies have collapsed before, and they will do so again. History is a databank of case studies in unsustainability.

Volumes have been written on the natural laws governing sustainability, and on the physical, economic, and social conditions for making sustainability real. Indicators of progress toward sustainability have been derived for cities, companies, and nations. What is sustainable, and what is not, is relatively well understood.

But it must be repeated: the word 'sustainability' is dying. It is dying because few concerted attempts have been made to enshrine a deeper understanding of the word in intellectual and political discourse, to defend the word from misappropriation, or to bring the word to public attention in a positive and exciting light. 'Sustainability' is dying of misuse, and dryness, and reduction to buzzword. It is dying because it is attached to too many initiatives that are failing to achieve their stated goals—or even, in many cases, to make any significant progress in that direction. It is dying because other initiatives, more cynically, pretend to be 'sustainable' when they are demonstrably not.

The misuses and abuses come from all sides. Sustainability is not a substitute word for environmentalism, though it is used as such by proponents and opponents alike. Sustainability is not a substitute word for economic growth, though it gets stretched in that direction far too often (as in 'sustainable growth'). Sustainable development—a term so misapplied as to be nearly beyond rescue—is not development-as-usual with a few green-looking additions or nods to social equity; but that is what it has often been reduced to in practice.

Sustainability is a far more ennobling concept than most current application reflects. Sustainability is a dream. Sustainability is an overarching ideal toward which any human society collectively strives. Sustainability is not 'the goal of all our striving', but it is the fundamental and primordial benchmark of our maturation as a species.

It is not an elegant word. It is, as words go, awkward, long, and technical in sound. But it is the best word we have for what we need: a vision. A direction. A set of criteria by which to measure our success.

Let us collectively abandon our use of the words sustainability and sustainable development, as they were used in the 20th Century. 'Sustainable development', in particular, has been abused almost beyond repair. Development—the change we make to the world—can either be good or bad. Good development contributes to sustainability; bad development makes sustainability more and more impossible, and collapse more and more certain. And most current development, including much of what is being done in the name of 'sustainable development', is quite bad, causing long-term damages far greater in scope than the benefits it purports to bring.

Let us therefore declare sustainability dead—and immediately proceed to revive it.

To be brought back to life, sustainability, as a word, must be reinvented. It must be imbued with all the qualities that our societies need to embrace to make sustainability itself possible. The word 'sustainability' should vibrate with creativity and shine with promise. Sustainability should fascinate the hungry mind, satisfy the heart in search of a meaningful life, draw people to it the way athletes are drawn to compete, the way artists are drawn to create, the way lovers are drawn to each other.

For our descendants, sustainability may someday be about maintaining a hard-won balance between the needs of people, nature's other species, and future generations of both. But we are far from balance today. For this generation, sustainability is about *global transformation*. Nothing could be more exciting to consider as the project of a generation, except perhaps making the first journey to the stars. We have before us the opportunity and the responsibility to begin *remaking our world*. We can, and we must, make it more beautiful in every respect, more delightful, more effective and efficient at securing our needs and encouraging our aspirations.

In the 21st Century, let us abandon diminished applications of this potentially enlightening word, and use 'sustainability' only when it carries the full radiance of a dream—the dream of civilization's transformation to a more uplifting, beautiful, ecological, equitable, and genuinely prosperous pattern of development.

The Transformation of Globalization

Transformation of many kinds is already happening all around us, mostly in the name of globalization. 'Globalization' has become the signifier for a family of transformations in communications, finance, trade, travel, ecological and cultural interaction that are drawing the world's people and natural systems into ever closer relationship with each other, regardless of national boundaries. Many of these transformations contribute more to the likelihood of global collapse than to global sustainability, because they are fueled by destructive technologies, they result in ever greater levels of environmental damage, they undermine national democracies, and they have so far widened dramatically the gap between rich and poor.

Yet there is nothing inherently unsustainable about globalization *per se*, if we understand that word to mean the growing integration of global human society. Indeed, globalization of many kinds—from the spread of better technologies to the universal adoption of human rights—is essential to attaining global sustainability. But the engines of globalization need to be harnessed to a more noble set of goals and aspirations.

At the heart of most descriptions of globalization is the market economy. It has often been fashionable to blame the market for the environmental crisis, and in particular to blame the market's tendency to concentrate power within large, independent capital structures we call 'corporations'.

But we need corporations, and the market, to accomplish the change we seek. To develop and spread innovations for sustainability at transformation speed, we need corporate-scale concentrations of research, production, and distribution capacity. We need the market's speed, freedom, and incentive structures. Clearly, we also need governors on the spread of destructive development, and the enormous fleet of old and dangerous innovations—from the internal combustion engine to the idea that cynical nihilism is 'cool'—that are increasing our distance from the dream of sustainability at an accelerating rate. But if we can alter globalization so that it turns the enormous power of the market and the corporation in a truly sustainable direction, we will watch in awe as our world changes for the better with unimaginable speed.

Envisioning the transformation of globalization will strike many as the ultimate in wishful thinking. Yet transformation begins precisely in wish and thought; and there are currently two powerful wishes adding considerable weight to global efforts to bring down the Berlin Wall between today's damaging 'capitalism-at-all-costs' and tomorrow's practice of a more mindful 'capitalism conscious of all costs'. One 'wish' is the new United Nations' 'Global Compact' with the corporate sector. It calls on corporations to adopt greater levels of social and environmental responsibility—a call that many are pledging to heed. The other 'wish' is the non-governmental Global Reporting Initiative, which sets new criteria for measuring sustainable corporate performance and is fast becoming adopted as the international standard, by corporations and activists alike.

These promising developments, still in their relative infancy, did not appear suddenly out of nowhere. There are but the latest and most successful demonstration of the power of 'wishful thinking', indulged in by hundreds of thousands of people, from the Seattle protesters of 1999 to the world government theorists of the 1930s. And these agreements are, themselves, 'wishful thinking' of a kind, comprised as they are of agreements on principle and criteria for measurements. But if this is what wishful thinking can do, consider what inspired action, multiplied throughout the global system, will accomplish when seriously embraced at the same scale.

Indeed, the transformation of globalization will, in many ways, signal the onset of transformation in general. When we witness the redirection of investment flows, the adoption of new rules and ethics governing the production process, the true raising of global standards of environmental, social, and economic performance, sustainability will then be written directly into the cultural genes, also known as 'memes', steering global development. These new 'sustainability memes' will then be replicated in every walk of industrial life. The dream of sustainability will become business as usual.

The Quest for Sustainability

We are still, however, quite a distance from that happy day. Moving decisively in the direction of sustainability will require transformative change in virtually every area of human endeavor. We must, at a minimum, accomplish the following:

- **Completely redesign and rebuild our energy systems** so that they drastically reduce carbon dioxide and other greenhouse emissions. The implications of this imperative are staggering: every internal combustion engine, every coal-fired power plant, every methane-emitting landfill must be transformed or replaced with an alternative that is climate-neutral and environmentally benign. We must speed up the innovation cycle and the depreciation cycle of capital investment. We need breakthroughs in the spread of solar, wind, hydrogen, and other forms of energy, together with new policies and financial instruments to accelerate the transformation process;

- Ideally, we should simultaneously **develop a globally coordinated system for managing the global carbon balance** at a scientifically determined acceptable level, since current best-case scenarios for emission reduction still leave us with an unacceptably warmer world;

- **Completely eliminate the threat of nuclear weapons and materials** from escaping into the biosphere. Highly radioactive and long-lasting materials like plutonium, especially, must be contained in perpetuity or transformed into more benign materials; and new technologies, both in science and in social patterns, must be discovered for achieving either goal;

- **Completely overhaul our production and use of chemicals** and materials so that no toxins of any kind are allowed to accumulate in the biosphere. A concerted effort is needed to identify existing alternatives, innovate new ones, and diffuse both throughout the global economy;

- **Eliminate global poverty and the threat of war.** 'Poverty reduction' is neither a noble nor an adequate goal, as poverty creates ecological destruction, increases social instability, and diminishes our humanity. War is too dangerous in an era of globally destructive weaponry. Nothing less than the full elimination of these two scourges is sufficient to attain sustainability and establish the full proof of our maturity as a species;

- **Protect absolutely the integrity of the Earth's natural and agricultural systems.** Hard boundaries should be drawn around biodiversity preserves, critical ecosystems, and places of awe and wonder. Farmlands and food production should be protected from displacement by urban sprawl and colonization by overzealous profiteers. Human habitations should be completely self-sufficient, no longer drawing down resources at unsustainable rates or destroying places of living mystery with thoughtless extraction, pollution, or overuse.

To achieve these and other lofty goals, change agents—people dedicated to promoting sustainability ideas and innovations—are needed in every field, in ever increasing numbers. We need, especially:

- The **artists**, to help us feel the gravity of our predicament, to facilitate our envisioning a more beautiful way of life, and to inspire us to strive for better things;

- The **scientists and engineers**, to find solutions, new inventions, break-through ideas that can rapidly transform our way of life;

- The **designers**, to redesign virtually everything, and to fuse beauty and functionality in a transformed world;

- The **business people**, to reimagine and redirect the flows of money and investment and talent in ways that can recreate the world while enhancing global prosperity;

- The **activists**, to call attention to those issues about which societies at large are in denial or unable to act because of systemic or hegemonic forces;

- The **professionals**, so-called, such as those in health care or the law or international development, to change the standards of practice in their profession and to lend their considerable weight to a general movement for change;

- The **average citizens**, so-called, to reimagine themselves as global citizens, to enthusiastically support change efforts, and to dare to reach for their own aspirations for a better world;

- The **politicians**, to motivate us with inspiring rhetoric, to frame new policies that encourage transformation, and to tear down obstacles to innovation and transformation;

- The **educators**, to prepare current and future generations for a great responsibility: directing human development toward sustainability, and beyond.

If a critical mass of people in all walks of life seriously take the charge to make transformation happen and if they are supported with widespread communication networks, resources, and incentives, then *transformation will happen*, and sustainability will become an attainable dream.

And *transformation will enrich us*, not impoverish us. It will enrich us spiritually, socially, and economically. We will know our purpose more profoundly, live together more compassionately, and develop wealth more equitably. There is so much work to be done that there will be jobs for all who will want them. So much genuine new value will be added to our economies that our measures of 'economic growth' will continue to rise, even as our impact on nature declines dramatically.

In fact, to achieve a genuine transformation, *we must accelerate and redirect our economies*, not slow them down. The demand for innovation, redesign, and redevelopment is too great to be achieved by anything less. Our responsibility for the dangers we have already created requires us to continue growing in our technical capacity, scientific understanding, and economic integration.

We can climb the mountain of sustainability, but not by pulling back. We must charge forward, and reach up, with all the strength, intelligence, wisdom, compassion, and determination of which our species is capable. And when we attain the summit, we will see the world from an entirely different perspective.

Life After Sustainability

We do not know, ultimately, what the purpose of life is, or even whether the concept of 'purpose' is a meaningful one. Our philosophical traditions provide a legacy of questions, but no ultimate answers. Our scientists can increasingly describe what the universe is and how it works, but they cannot approach the ultimate question of why. Our religious traditions, in all their diversity, do approach this question, and they provide hints and guidance and, for some, the solace and foundation of faith—but the ultimate unanswerability of life's greatest questions is precisely the reason for religion's existence.

Attaining sustainability does not release humanity from wrestling with such questions as, Where do we come from? Why do we suffer and die? How shall we live?

But the closer we get to sustainability, the more we can address these questions in full freedom—and the more our descendants will be free to consider them, unburdened by poverty, or ecological instability, or insecurity about the future of civilization. We do not know what a sustainable world will look like, but we can be assured that it will far more beautiful, creative, prosperous, fascinating, and engaging of our full humanity than the world in which we now live.

The challenges are enormous, and the indications of success are largely visible only over the course of years or decades. But the rewards, even for making the attempt, are great—for all of us now, and with luck, for all the generations of life to come after us.

About the Authors

ALAN ATKISSON

Alan AtKisson is Founder and President of the AtKisson Group, an international consulting network focused on sustainable development, with representatives and licensees in ten countries.

He is the author of the popular 1999 book, *Believing Cassandra: An Optimist Looks at a Pessimist's World*. The book was a bestseller on Amazon.com in the US, and has also been published in both Australian and Japanese editions. He is also a co-author of *The Natural Advantage of Nations*, a comprehensive review of sustainability in business and governance edited by Michael Smith and Karlson Hargroves, published by Earthscan in September 2004.

Since founding AtKisson Inc. in 1992, AtKisson has consulted to well over one hundred cities, companies, government agencies, civic initiatives, and non-profit organizations, and he has presented hundreds of keynote speeches, seminars, workshops and training sessions worldwide.

AtKisson began his work in sustainability as editor of *In Context*, a pioneering journal of sustainable cultures and systems, and as co-founder of Sustainable Seattle, a civic engagement initiative, which introduced a model for urban sustainability indicators and was recognized with awards by several organizations including the United Nations. He is a former Senior Fellow and Executive Director of the independent policy institute Redefining Progress. He is also a founding member of the International Consultative Group on Sustainable Development Indicators as well as the International Sustainability Indicators Network.

The Earth Charter Steering Committee, which was given responsibility for oversight and direction of the Earth Charter Initiative (ECI) in 2000 by the Earth Charter Commission, appointed Alan AtKisson to the newly created position of International Transition Director in the Fall of 2005. In January 2006, Mr. AtKisson assumed his new responsibilities as the chief executive officer of the Earth Charter Initiative.

ALBERT A. BARTLETT

Albert Bartlett is a retired Professor of Physics who joined the faculty of the University of Colorado in Boulder in September 1950. His BA degree in physics is from Colgate University and his MA and PhD degrees in physics are from Harvard University. In 1978 he was national president of the American Association of Physics Teachers. He is a Fellow of the American Physical Society and the American Association for the Advancement of Science. In 1969 and 1970 he was the elected Chair of the four-campus Faculty Council of the University of Colorado.

In the late 1950s Bartlett was an initiator of a citizens' effort to preserve open space in Boulder, and this ultimately led to the establishment of the City of Boulder's Open Space Program (1968), which has purchased over 44,000 acres of land to be preserved as public open space. He is a founding member of PLAN-Boulder County, an environmental group for the city and county.

Since the late 1960s he has concentrated on public education on problems relating to and originating from population growth. Since 1969 he has given his lecture, *Arithmetic, Population, and Energy* 1,580 times to audiences of all levels across North America. More recently he has concentrated on writing on sustainability, examining the widespread misuse of the term and the conditions that are necessary for attaining sustainability in any society.

MARIOS CAMHIS

Marios Camhis studied at the National Technical University of Athens and the Planning Department of the Architectural Association in London. He obtained his PhD at the London School of Economics followed by post-doctoral studies at UCLA. His work was published in book form in 1979 under *Planning Theory and Philosophy* (Tavistock Publications). He has lectured at several universities and has written articles on Urban and Regional Policy and European Affairs. He is also an Honorary Member of the Royal Town Planning Institute (UK).

Camhis has worked as Advisor at the Ministries of National Economy and Foreign Affairs in Greece. Since 1985 he is a European Commission official. Among other duties, he has headed a division responsible for urban issues, spatial planning and cross-border cooperation in the Directorate General of Regional Policy. Between 1997 and 2001 he served as Director of the Representation of the European Commission in Greece. He spent the academic year 2002/2003 as a European Union Fellow at the School of

Public Policy of George Mason University, teaching Regional and Sustainable Development Policies. He is currently Advisor to the Directorate General of Press and Communication of the European Commission in Brussels.

HERMAN E. DALY

Herman E. Daly came to the Maryland School of Public Affairs from the World Bank, where he was Senior Economist in the Environment Department, helping to develop policy guidelines related to sustainable development. At this post, he was primarily engaged in environmental operations work in Latin America.

Before joining the World Bank, Daly was Alumni Professor of Economics at Louisiana State University. He is a co-founder and associate editor of the journal *Ecological Economics*. His interest in economic development, population, resources, and the environment has resulted in numerous articles as well as books, including *Steady-State Economics* (1977; 1991), *Valuing the Earth* (1993), *Beyond Growth* (1996), and *Ecological Economics and the Ecology of Economics* (1999). He is co-author with theologian John B. Cobb, Jr. of *For the Common Good* (1989; 1994), which received the Grawemeyer Award for ideas for improving world order.

He is a recipient of the Honorary Right Livelihood Award (Sweden's alternative to the Nobel Prize), the Heineken Prize for Environmental Science from the Royal Netherlands Academy of Arts and Sciences, and the Sophie Prize (Norway).

MIKHAIL GORBACHEV

Mikhail Gorbachev was born on March 2, 1931 in the village of Privolnoye in the south of the Russian republic.

The experience of the Nazi Germany invasion and the Great Patriotic War left a lasting impression in his mind. Still a schoolboy, Mikhail started working, helping his father operate a combine harvester. For outstanding results in bringing in the bumper crop Gorbachev was awarded the Order of the Red Banner of Labor. He was only seventeen then and became the youngest recipient of this high award.

In 1950 Gorbachev was admitted to Moscow State University. He studied at the law faculty, graduating in 1955. Later, he took correspondence courses from Stavropol Agricultural Institute, and in 1967 added a degree in agricultural economics to his Moscow law degree.

As a student at the University Gorbachev joined the Communist Party of the Soviet Union. From his youth he had been active in public life and social and political causes, eager to find solutions to problems. He was elected to leadership positions whenever and wherever he worked.

The university years were prominent in shaping the mind of Mikhail Gorbachev. Also at that time, Mikhail Gorbachev married Raisa Titarenko. Since then and till the end of her days, Raisa Gorbachev was the person most close and dear to Mikhail Gorbachev wherever he worked, no matter in what capacity. She passed away in September 1999.

Having received his degree, Gorbachev was ready to work as a lawyer. At first he was offered a job in the Prosecutor's Office Moscow, then in the Stavropol Territory. Soon upon his return to the home city of Stavropol, however, he was offered a position in the local Komsomol youth league. Thus his political career started.

From 1955 to 1962 he rose fast through a variety of Komsomol jobs, winding up as the First Secretary, Stavropol Territory Komsomol Committee, and then was given a party job. In 1966 he was elected First Secretary, Stavropol City Party Committee. In August 1968 he became Secretary, Stavropol Territory Party Committee, and in April 1970 First Secretary, the highest post in Stavropol Territory. In 1970 he was elected member of the CPSU Central Committee. In November 1978 he became a Central Committee Secretary and moved to Moscow. Two years later he joined the Politburo of the CPSU Central Committee. At a time of stagnation, many people perceived his fast rise as a sign of imminent changes in Soviet society. In March 1985, Gorbachev was elected General Secretary of the Party Central Committee.

Gorbachev initiated the process of change in the Soviet Union - what was later called perestroika, the fundamental transformation of the nation and society. Glasnost became perestroika's driving force. A sweeping process of the nation's democratization was launched and reforms were planned to put the nation's ineffective economy back on track to market economics.

A big shift in international affairs was effected. The new thinking associated with the name of Gorbachev contributed to a fundamental change in the international environment and played a prominent role in ending the Cold War, stopping the arms race and eradicating the threat of a nuclear war.

In 1988, Gorbachev became Chairman of the Presidium of the Supreme Soviet of the USSR, and in 1989, Chairman of the Supreme Soviet of the USSR. Accordingly, he headed the nation's Defense Council and was Supreme Commander-in-Chief of the Armed Forces of the USSR. The Congress of People's Deputies of the USSR elected Gorbachev President of the USSR on March 15, 1990. On December 25, 1991, Gorbachev stepped down as Head of State.

In recognition of his outstanding services as a great reformer and world political leader, who greatly contributed in changing for the better the very nature of world development, Mikhail Gorbachev was awarded the Nobel Peace Prize in 1990.

Since January 1992, he has been President of the International Nongovernmental Foundation for Socio-Economic and Political Studies (The Gorbachev Foundation). Since March 1993, he has also been President of Green Cross International—an international independent environmental organization with branches in almost thirty countries.

Marco Keiner

Marco Keiner studied geography at the University of Erlangen-Nuremberg and spatial planning at the Swiss Federal Institute of Technology (ETH) in Zurich. In 1999, he received his PhD in geography from the University of Eichstätt with a dissertation on sustainable rural development in Mali. In 2005, he acquired his 'Habilitation' (postdoctoral lecturing qualification for German-speaking universities) at the University of Innsbruck, based on a thesis on new and improved sustainability-oriented regional planning instruments.

Keiner is senior researcher, postgraduate lecturer and head of the Sustainability Research Group of the Institute for Spatial and Landscape Planning at ETH Zurich and has worked extensively in developing countries, and in the European context. In 2004, he coined the concept of *Evolutionability*, thus re-centering the sustainability discussion on environmental and ethical issues. He recently co-edited the books *The Real and Virtual World of Spatial Planning* (Springer, 2003), *From Understanding to Action—Sustainable Urban Development in Medium-Sized Cities in Africa and Latin America* (Springer, 2004), and *Managing Urban Futures—Sustainability and Urban Growth in Developing Countries* (Ashgate, 2005).

Klaus M. Leisinger

Klaus M. Leisinger is President and CEO of the Novartis Foundation for Sustainable Development and Professor for Development Sociology at the University of Basel.

Leisinger studied economics and social sciences at the University of Basel, Switzerland. His postdoctoral studies included field and desk research on *Health Policy for Least Developed Countries*, population policy, and international business ethics. His professional career included a four-year term as CEO of the former Ciba Pharmaceuticals regional office in East-Africa. Since July 1, 1990, Leisinger has been Executive Director and Delegate of the Board of Trustees of the Novartis Foundation for Sustainable Development. Since July 2002, he also serves as President of the Foundation.

In addition to his work at Novartis, he teaches and conducts research as Professor of Sociology at the University of Basel. Focus themes include Business Ethics and Globalization, Corporate Social Responsibility of Pharmaceutical Corporations, as well as Human Rights and Business. He has contributed to debates on these issues with a significant number of publications in several languages. Leisinger has been an invited lecturer at several Swiss and German universities as well as Georgetown University, the University of Notre Dame, MIT Sloan School of Management, and Harvard University.

Klaus Leisinger has held advisory positions in a number of national and international organizations, such as the United Nations Global Compact, the United Nations Development Program (UNDP), the World Bank (CGIAR), Asian Development Bank as well as Economic Commission for Latin America (ECLA). Among others, he is member of the Board of Trustees of the German Network Business Ethics and of the Advisory Council of Mary Robinson's Ethical Globalization Initiative. In September 2005, Kofi Annan appointed Klaus Leisinger Special Advisor of the United Nations Secretary General for the UN Global Compact.

PETER MARCUSE

Peter Marcuse, a lawyer and urban planner, is Professor of Urban Planning at Columbia University in New York City. He has been involved with urban policy for many years and served as the Majority Leader of Waterbury, Connecticut's City Council and member of its City Planning Commission. Later he was President of the Los Angeles Planning Commission and more recently, a member of Community Board 9 in Manhattan as well as co-chair of its Housing Committee. He ran a private law practice for over 20 years before becoming Professor of Urban Planning, first at the University of California at Los Angeles, from 1972 to 1975, and then at Columbia University. He has taught in Germany (West and East), Australia, Canada, Austria, Brazil, Hungary, and South Africa. Since 2003 he is semi-retired, with a reduced teaching load.

Marcuse has long-standing interests in globalization, comparative housing and planning policies. *Globalizing Cities: A New Spatial Order of Cities* (Blackwell, 1999), co-edited with Ronald van Kempen, deals with the impact of globalization on the internal urban structure of a diverse set of cities around the world. Marcuse's newest book, also co-edited with van Kempen, *Of States and Cities* (Oxford University Press, 2002), looks at the role of governments in urban development. He is currently working on a book on the history of working-class housing in New York City. Marcuse has written widely on social housing, housing policy, red-lining, racial segregation, urban divisions and the dual and quartered city, New York City's planning history, legal and social aspects of property rights and privatization, the transition from 'socialism' in eastern Europe, professional ethics, and the history of housing. He also spoke at both meetings of the World Social Forum in Porto Alegre, Brazil. He was an early member of the Planners Network, an organization of progressive planners in the United States and remains active in its efforts to influence planning in New York City in the aftermath of September 11 in a manner that promotes equity and social justice.

Marcuse is on the editorial boards of a number of professional journals and has acted as consultant to local, state, and national governments on housing policy issues. The meaning and impact of globalization on housing and urban social spatial patterns within a comparative perspective, with a focus on social justice, is the main theme of his current work. He has also written on the impact of September 11 on New York City and on globalization, focusing on the attack's impact on social justice.

DENNIS L. MEADOWS

Dennis Meadows is President of the Laboratory for Interactive Learning and Emeritus Professor of Systems Policy and Social Science Research, University of New Hampshire, where he was formerly Director of the Institute for Policy and Social Science Research. Between 1969 and 1988, Meadows was on the faculty of the MIT System Dynamics Group and Dartmouth College's Tuck School of Business and Thayer School of Engineering. As a young faculty member he took a year's leave of absence from MIT and drove from London to Sri Lanka and back. That trip gave him the incentive and many useful insights for a lifelong career focused on the interaction of economy, politics, ecology, technology, and culture as they influence long-term patterns of development in human societies.

As author and co-author, Meadows has written ten books on aspects of forecasting, futures, educational games, systems analysis, sustainable development, social change, and the global economy. The books have been translated into more than 30 languages. Included is the most widely sold computer simulation book in history, *The Limits to Growth.*

Meadows is President of the Balaton Group, a network of 100 professionals in 40 nations involved in systems science and public policy, cofounder of the Resource Systems Group, and Past President of the International System Dynamics Society and the International Simulation and Games Association. He has served as corporate board member and consultant for government, industry and non-profit groups in the US and many countries abroad. He has a PhD in Management from MIT, where he served on the faculty, and three honorary doctorates. Meadows also participated for ten years in the management board of The Browne Center, where he facilitated workshops and contributed innovative and complex games to the curriculum, including sophisticated games to convey principles of systems thinking for managers in corporate and public sectors.

HELENA NORBERG-HODGE

Helena Norberg Hodge is the founder and Director of the International Society for Ecology and Culture (www.isec.org), a non-profit organization concerned with the protection of both biological and cultural diversity and 'education for action' or moving beyond single issues to look at the more fundamental influences that shape our lives. She is also a board member of the International Forum on Globalization (www.ifg.org), an alliance of sixty leading activists, scholars, economists, researchers, and writers formed to stimulate new thinking, joint activity, and public education in response to economic globalization.

Norberg-Hodge focuses on examining and critiquing conventional notions of development. A linguist by training and a native of Sweden, she was educated in Europe and the US. She first went to Ladakh in northwestern India in 1975. Three years later she founded the Ladakh Project, with the goal of providing Ladakhis with the means to make more informed choices about their own future. Her work to support traditional societies endangered by globalization has received wide support and recognition. She is also the author of several books, including *Ancient Futures: Learning from Ladakh.*

In 1986, Helena Norberg Hodge received the Right Livelihood Award (Alternative Nobel Prize).

MATHIS WACKERNAGEL

Mathis Wackernagel, PhD, is a founder and Executive Director of Global Footprint Network, a California-based non-profit that supports a sustainable economy by using the Ecological Footprint to make ecological limits central to decision-making everywhere. Wackernagel has worked on sustainability issues for organizations in Europe, Latin America, North America, Asia and Australia. He has lectured for community groups, governments and their agencies, NGOs, and academic audiences at more than 100 universities on all continents but Antarctica.

Wackernagel has authored or contributed to over fifty peer-reviewed papers, numerous articles and reports, and various books on sustainability that focus on the question of embracing limits and developing metrics for sustainability, including *Our Ecological Footprint: Reducing Human Impact on the Earth, Sharing Nature's Interest*, and WWF International's *Living Planet Report*. After earning a degree in mechanical engineering from the Swiss Federal Institute of Technology, he completed his Ph.D. in community and regional planning at The University of British Columbia in Vancouver, Canada. There he created, as his doctoral dissertation, with Professor William Rees the 'Ecological Footprint' concept, now a widely used sustainability measure. Wackernagel is also an adjunct faculty at SAGE of the University of Wisconsin-Madison, and scientific advisor of the Centre for Sustainability Studies in Veracruz, Mexico.

ERNST ULRICH VON WEIZSÄCKER

Ernst Ulrich von Weizsäcker was Professor for Interdisciplinary Biology at Essen University (1972–1975), Founding President of the University of Kassel (1975–1980), Director at the United Nations Centre for Science and Technology for Development (1981–1984), Director of the Institute for European Environmental Policy (IEEP) in Bonn, London, and Paris (1984–1991), and President of the Wuppertal Institute for Climate, Environment, and Energy (1991–2000). Since 2006, he is Dean of the Donald Bren School of Environmental Science and Management at the University of California at Santa Barbara.

Since 1991, von Weizsäcker has been a member of the Club of Rome, and from 1998–2005 member of the German Parliament and Chairman of the Bundestag Committee on Environment, Nature Conservation, and Nuclear Safety. He was Chairman of the Parliamentary Commission on the

Enquiry 'Globalization of the Economy—Challenges and Responses' and-
from 2002 to 2004 member of the World Commission on the Social Dimension of Globalization.

Von Weizsäcker has received numerous awards, including the Italian
Premio De Natura together with the Norwegian Prime Minister Gro
Brundtland, the Duke of Edinburgh Gold Medal of WWF International, an
Honorary Degree from Soka University in Japan, and the Takeda Award
'Techno-Entrepreneurial Achievement for World-Environmental Well-
Being' together with Prof. Dr. Friedrich Schmidt-Bleek.

Index

Printed in the United States
64695LVS00002B/190-210

9 781402 047343